Praise for Brad Matsen's

DESCENT

"Fascinating. . . . The pages explaining the descent in detail are so vivid, I felt like I was holding my breath. . . . Matsen braids it all together in dramatic detail."

—Patti Ross, *San Antonio Express-News*

"Explorers everywhere will rejoice over *Descent*, Brad Matsen's thoughtful, engaging, and entertaining account of ocean exploration. And who is not an explorer at heart?"

—Sylvia A. Earle, National Geographic
Explorer-in-Residence and author of
Sea Change: A Message of the Oceans

"Matsen tells [the story] very well, with a surprising degree of suspense." —*Discover*

"I could not put this book down; *Descent* is fascinating not only as a tale of high scientific adventure but also as a psychological study of two men who risked their lives to plumb the abyss." —Oliver Sacks, author of *Uncle Tungsten*

"As exciting as any fictitious adventure. The story speeds along at a lively pace with equal parts suspense, conflict and exhilaration. . . . Matsen tells his story with depth and clarity."

—*Winston-Salem Journal*

"Suspenseful. . . . *Descent* skillfully relates how two men hazarded their all to see the wonders of the deep; this is a testament to what can be achieved when unflagging curiosity is fueled by mad courage."   —Caroline Alexander, author of *The Endurance* and *The Bounty*

"Well-written, engaging."   —*The Seattle Times*

"With his trusty sidekick Otis Barton, William Beebe prepares to descend in a sealed steel ball to a half mile down, to look for weird and wonderful deep-sea creatures. Will he find them? Will he and Barton succeed in their record-breaking dive? In this wonderful book, you will find the answers to these and any number of other fascinating questions about deep-sea exploration."
                    —Richard Ellis, author of *Monsters of the Sea*

"Riveting. . . . [The bathysphere's] story, as told in these pages, is every bit as interesting as the stories of Beebe and Barton. This book will appeal to anyone interested in adventure or marine exploration."   —*Science News*

"Gripping."   —*Tucson Citizen*

"Brad Matsen's *Descent* is the *Seabiscuit* of the deep ocean. It evokes an amazing era when we humans were just discovering the planet we live on, and it never lets up with the drama of risk-taking in the name of science. I absolutely loved this book." —Ray Troll, artist and author of *Rapture of the Deep*

Brad Matsen

DESCENT

Brad Matsen is the author of many books about the sea and its inhabitants. He was a creative producer for the television series *The Shape of Life,* and his articles on marine science and environmental topics have appeared in *Mother Jones, Audubon, Natural History,* and other magazines and anthologies. He divides his time between Seattle and New York City.

DESCENT

The Heroic Discovery of the Abyss

Brad Matsen

VINTAGE BOOKS

A Division of Random House, Inc.

New York

FIRST VINTAGE BOOKS EDITION, JUNE 2006

Copyright © 2005 by Brad Matsen

All rights reserved. Published in the United States by Vintage Books, a division of Random House, Inc., New York, and in Canada by Random House of Canada Limited, Toronto. Originally published in hardcover in the United States by Pantheon Books, a division of Random House, Inc., New York, in 2005.

Grateful acknowledgment is made to the following for permission to reprint previously unpublished material: James H. Barton: Excerpted material from Otis Barton's correspondence and papers; excerpts from the unpublished memoir written by Francis L. Barton. Reprinted by permission of James H. Barton. • Bermuda Biological Station for Research: Excerpts from papers and correspondence on the Bathysphere expeditions. Reprinted by permission of the Bermuda Biological Station for Research. Princeton University Library: Excerpted material from the papers of William Beebe. Princeton University Library, Manuscripts Division, Department of Rare Books. Reprinted by permission of Princeton University Library. • Wildlife Conservation Society: Excerpts from the letters and reports written by William Beebe, John Tee-Van, Gloria Hollister, and Jocelyn Crane. Reprinted by permission of the Wildlife Conservation Society.

The Library of Congress has cataloged the Pantheon edition as follows:
Matsen, Bradford.
Descent : the heroic discovery of the abyss / Brad Matsen.
p. cm.
Includes bibliographical references and index.
1. Underwater exploration. 2. Beebe, William, 1877–1962. 3. Barton, Otis. I. Title.
GC65.M365 2005
551.46'092'2—dc22
2004053530

Vintage ISBN-10: 1-4000-7501-7
Vintage ISBN-13: 978-1-4000-7501-0

Author photograph © Jonas Bendiksen/Magnum
Book design by M. Kristen Bearse

www.vintagebooks.com

Printed in the United States of America
10 9 8 7 6 5 4 3 2 1

FOR MILO

CONTENTS

ILLUSTRATIONS

All © Wildlife Conservation Society, headquartered at the Bronx Zoo, except where otherwise noted.

CONEY ISLAND

*You can't begin to understand the universe
unless you look at it once in a while.*

John Dobson, astronomer

The directions came from a friend who had gotten them from
another friend, but as expeditions go, it was cheap and easy. For
a dollar fifty, I rode the F train from Manhattan over the graffitied
rooftops of Brooklyn to the Avenue X station, caught a shuttle bus to
Surf Avenue, and walked between the Cyclone roller coaster and Paul
and Gregory's hot dog stand. The aroma of sauerkraut on the summer
breeze was delicious. I stopped to listen to the *racketa-racketa* of the
climbing Cyclone and the screams of its riders enjoying a heart-stopping
plunge. Then I shuffled up a sandy concrete ramp to the boardwalk and
gazed at the Atlantic, which spread out before me. The sea was glassy
calm, its steel-gray sheen studded with the bright dots of swimmers bobbing between the rock fingers of the jetties. In the middle distance, the
blues of deeper water took over from the lighter tones of the shallows
and carried my eye to the horizon, beyond which I imagined Europe and
Africa, the brave ancient sailors who dared to leave the sight of land far
behind, thousands of shipwrecks, gales, riots of stars overhead on pitch-
dark nights, and an entire universe hidden below the water's surface
about which we know only a bit more than we do about Mars.

I walked past a weathered pastel mural of King Neptune in a chariot
pulled by sea horses and under an arch to the main gate of the New York
Aquarium. The entrance plaza was crowded with some of the million or
so people who line up there every year to look at real sea horses, sharks,
jellyfish, and the other inhabitants of the wild, hidden realm that occu-
pies more than two-thirds of the earth's surface. I doubled back down a

hedge-lined path under an elevated walkway and into a parking lot, keeping my eyes peeled for security patrols, because once I left the parking lot I'd be trespassing.

The coast was clear, though, and ahead of me I saw a storage yard right beneath the end of the roller coaster nearest the sea. It was surrounded on three sides by chain-link fence but open to the parking lot and unguarded. I made my way over some long sections of PVC pipe scattered like fallen pickup sticks, past two six-foot-high piles of wooden pallets, two stacks of ordinary glass fish tanks, and several columns of nested plastic flowerpots. Against the fence, with the worn, white trestles of the Cyclone towering directly over it, was a steel globe just about shoulder high. Several coats of paint had peeled successively, leaving a mottle of light blue, dark blue, and white, and on the side I saw painted in faded letters:

NEW YORK
ZOOLOGICAL SOCIETY

BATHYSPHERE

NATIONAL
GEOGRAPHIC
SOCIETY

Finding the Bathysphere in an outdoor scrap yard under the Cyclone in Coney Island was like coming across a Mercury space capsule among rusting tools and nicked furniture at a flea market. In it, William Beebe and Otis Barton were lowered on a cable into the Atlantic Ocean off Bermuda sixteen times between June 1930 and August 1934. Before them, no one had ever gone more than 350 feet into the ocean's depths and lived to tell about it, but they went to a depth of three thousand feet. I squeezed past the columns of flowerpots, stumbled when a pallet plank gave way, and bent to look at the tiny windows through which Beebe and Barton had gazed into the depths. They were the first human beings to descend beyond sunlight to a world of eternal darkness, inhabited by creatures out of nightmares, predators emblazoned with colored lights whose survival depends on frightening tactics of allure and deception. Reports of their sensational exploits made Beebe and Barton international celebrities during the Depression, when people craved vicarious thrills, and their scientific discoveries and courage inspired generations of oceanographers and deep-ocean explorers.

On the other side of the sphere, at the hatch, which was covered by a piece of clear plastic, I bent down and squared my shoulders to see if they would fit. It was obvious that even someone of my medium build would have to contort to board the Bathysphere, and it didn't look like there was room for a large dog inside, let alone two men. I measured the Bathysphere's diameter by standing dead center, putting my elbows to my sides, and extending my arms and hands to right and left. About four and a half feet. I looked up at the lifting lug on top, where a stub of cable remained, and with my thumb gauged the thickness of the filament upon which their lives had depended. About an inch. I imagined the wire rising fifty feet to the highest rail of the roller coaster and then sixty times beyond, farther into the sky than two Empire State Buildings stacked on top of each other. I thought of Beebe and Barton huddled shoulder to shoulder a half mile beneath the surface of the sea, shivering with cold and fear, but finding within themselves whatever is common to heroes and pioneers that enables them to risk death to explore the unknown.

As the Cyclone with its cargo of screamers clattered above me, I closed my eyes and rested my palms on the warm steel of the Bathysphere. Like anyone with even a passing interest in the ocean, I knew a few facts about William Beebe and Otis Barton, but I had no idea how those two men had summoned the courage to squeeze into this crude little craft to become the first human beings to descend into the abyss.

Part One

SURVIVAL

One

BARTON

If his inmost heart could have been laid open,
there would have been discovered that dream of undying fame;
which, dream as it is, is more powerful than a thousand realities.

Nathaniel Hawthorne, *Fanshawe*

When Otis Barton told the story, he always said it began and very nearly ended on Thanksgiving Day in 1926, when he went for a walk to buy a newspaper. He left his third-floor apartment on East Sixty-seventh Street in Manhattan and turned toward Madison Avenue, loping along lost in thought. Barton was preoccupied that morning with a recurring fantasy in which he was a celebrated explorer just back from a dangerous adventure, with photographs and specimens of creatures never before seen by man, resting between expeditions in a penthouse apartment with a weeping willow on the terrace, a beautiful girl who liked camping and looked good in a pith helmet by his side. Like other boys enchanted by the novels of Arthur Conan Doyle, Jules Verne, and the other fantasists who were so popular during his childhood, Barton had feasted on daydreams of wild animals, caves filled with gold, and lost civilizations. As a young man of twenty-six, the theater of his imagination was still as vivid to him as the pavement beneath his feet.

In the real world, though, Otis Barton was an engineering student at Columbia University, the grandson of a merchant who had started with a clapboard storefront in Manchester, New Hampshire, in 1850, sold dry goods, and prospered. Otis Barton's father, Frederick, went to Harvard, made a small fortune as a textile mill salesman during the boom years just before the turn of the century, and moved his family to New York, where business was even better. His first son, Frederick Otis Barton, Jr., was born there on June 5, 1899, followed by two daughters, Ellen and

Mary, and a second son, Francis. Frederick Barton died suddenly in 1905—heart attacks ran in the family—and his wife, the former Mary Lowell Coolidge, packed up her children and moved first to Concord, Massachusetts, and then to a house on Marlborough Street in Boston.

Mary Barton's relatives and social circle included the Lowells, Cabots, and Coolidges, who had built their fortunes in railroads, manufacturing, and finance as coal, steam, and cheap labor drove the engines of the Industrial Revolution. Her marriage to Frederick Barton had been considered shrewd by family patriarchs, who correctly assumed that merchants would capture a significant share of the money pouring into the pockets of mill and factory workers as mass consumption became a predictable part of the economic equation. After Mary returned to Boston as a widow with her share of the Barton inheritance and her own small fortune, she settled her family into the proven aristocratic pattern of fall and winter in the city, spring traveling abroad, and summers on Vineyard Sound.

Otis Barton had been raised by women. His mother, her sisters, and a devoted nurse, Katy Gaule, tended him like a prince, and the only indelible image of men in his life was a painting by William Merritt Chase that hung in their parlor entitled *A Portrait of Master Otis Barton and His Grandfather.* Against the kind of dark background familiar to anyone who has wandered through a gallery of Dutch masters stands the figure of four-year-old Otis in a high-collared, thigh-length smock and knee socks, next to a seated, gray-bearded patrician holding a sheaf of papers on his lap. Otis, looking directly at the artist, is a beautiful child with an oval face, dark hair combed to a shock in the middle of his forehead, and perfectly spaced features that hint of intelligence. His grandfather's countenance is tragic by comparison, defined by sagging pouches under his eyes and an expression of utter weariness. The painting seems to suggest sadly that all lives pass from hope to defeat.

At twelve, Otis Barton joined the tribe of privileged teenaged boys at Groton, where he played baseball and became something of a legend because of his academic record. Before he could read or write, he had discovered that he could think in pictures and recall images in his mind as though he were looking at a photograph or a painting. Otis used this rare gift of an eidetic memory as a parlor trick, reciting long passages in Greek and Latin after seeing them just once on a page. He could do the same thing with figures and reading assignments. Rote learning and memorization were in vogue, and he scored the second-highest grades in the history of the school up to that time. Otis was a tall, good-looking young man who should have fit in well with his classmates, but he often came across as moody and awkward, perhaps because the terrain of his imagination was every bit as real to him as that of the outside world. By the time he left Groton he was known as a loner and a daydreamer.

Real adventure broke through Otis Barton's solitary fantasy life most often during summers in Cotuit on Vineyard Sound, where his family had a mansion they called a cottage, an enormous seaside pile of dozens of small rooms and porches with lawns sloping to the sea. The house was divided into a women's wing and a men's wing shared by Otis and Francis with a steady stream of tutors and hired playmates. The Bartons devoted their summers to picnics, swimming, boating, outings to neighboring towns, ice cream socials, and costume parties. Sometimes Otis holed up alone, reading or simply lying in his bedroom or on a cot on the porch, getting up only for meals, but on other days he organized energetic adventures. One summer, when he was reading *In Darkest Africa*, Henry M. Stanley's account of becoming the first European to cross the Dark Continent, Otis became obsessed with the way the natives captured animals in pit traps. He persuaded his brother to help him dig one outside the toolshed, cover the pit with branches, seaweed, and sand, and trap their gardener, George Childs. The story endured decades of repetition at family gatherings.

The ocean, though, was the dominant presence of Otis's summers in Cotuit. In sailing dinghies, he and Francis would follow the gray-blue forms of sand sharks over the shoals of Vineyard Sound and sometimes spear them for sport. Once, Otis dove at one of the shadows from the crosstrees of a large sailboat with a knife in his hands, but the shark was too fast for him and got away unharmed. Otis burned with curiosity about the realm of shark shadows and unseen treasures and demons beneath the sea, and during his summers at Cotuit he recapitulated the history of human attempts to descend into its mystery. After he saw a drawing of a naval battle between the Greeks and the Syracusans in which saboteurs swam invisibly underwater by breathing through hollow reeds, he took a length of garden hose and, with Francis holding the end of the hose in the air, weighed himself down with bags of BB shot, held his nose, and walked along the sloping bottom from the beach, taking sips of air from the hose. At a depth of about six feet, he could no longer draw fresh air through the tube, and he realized that his lungs just weren't powerful enough to pull the air that far from the surface. Obviously, Otis thought, the only way to go deeper was to bring air with him. He had seen an etching of Alexander the Great sitting on the bottom of the sea under a barrel, so he tried a dive with a washtub over his head, tied by its handles to his shoulders. He could breathe, but the air made his washtub so buoyant he couldn't sink more than a few feet, even with weights.

During the summer of his sixteenth year, Barton's passion for sailing, shark hunting, and fishing was enriched after he found a diving equipment catalog at a boatyard in Cotuit. The catalog advertised a selection of professional gear available in 1915 that men used for salvage work and for exploring to depths of sixty feet. A complete outfit with helmet and suit was too expensive, but he ordered a small brass pump that could send air down to a depth of thirty feet and a length of nonkinking hose. The brass and copper helmets in the catalog were sturdy-looking, with molded shoulder plates. The concept was simple, so Barton sketched out a plain wooden box with a glass pane in front and a hose coupling on top with straps on the sides that ran under his armpits, and took his design to a cabinetmaker in Boston, who built it for next to nothing.

When the helmet arrived on the freight wagon, he and Francis immediately hauled it, the pump, and the hose to the Cotuit town wharf. While a small crowd gathered, Barton attached bags of BB shot to his belt, slid into the arm straps, and lifted the helmet onto his shoulders. If he got into trouble, his plan was to release his weight belt and let the air in his helmet bring him to the surface. With Francis and another boy manning the pump and hose, Otis climbed down the wharf ladder and settled to the mud of the harbor twenty feet down. But a minute later he reappeared, shinnying back up a piling and heaving himself up the ladder. Between the buoyancy of the helmet and its armpit straps acting to lift him up and his weights holding him down, he was being torn in two. So Barton hung the shot from the helmet instead of his waist belt. This was more dangerous because he couldn't easily get rid of the weight in an emergency, but it was the only way to stay comfortably submerged while wearing a light wooden helmet full of air. If he couldn't breathe, he would just take off the helmet and hope for the best.

Barton climbed back down the ladder and this time walked around on the bottom of Cotuit Harbor for half an hour. The rhythmic panting of the pump sounded loudly in his ears, and he was surrounded by a dim light that he would later describe to Francis as "church-like." The murky water of the heavily used boat basin surrendered no great wonders, but Barton thought the glimpses of debris, old moorings, eel pots, a few shrimp, and the odd flatfish skipping away from his bare feet were miraculous. Until the end of the season in August, Barton explored in his helmet almost every day and often manned the pump while Francis and the other children explored the world beneath the sea. And then it was back to Boston.

In September, Barton stuck with family tradition and went to Harvard College, where he worked his way through courses in engineering, mathematics, and natural science with the same lackluster ambition but spectacular results that had worked for him at Groton. After his graduation in the spring of 1922, he took off on the trip around the world that was almost obligatory among the young men of his social class. Big game hunting was the rage, and he spent a few months shooting lions, tigers, elephants, and antelope on the African savanna and in the jungles of India. When he got tired of roughing it, he meandered eastward along the chain of deluxe colonial hotels in exotic locales and eventually fetched up on the Sulu Sea in the Philippines, where native divers told him about battles with enormous octopi and about giant clams with pearls as big as fists just beyond the reach of a man holding his breath underwater. Barton spent hours in the shallows of the tropical ocean, mesmerized by the gaudy reef fish and splashy corals and tantalized by the deep water, where the color shifted from aquamarine to deep blue to purple.

Back in New York with his wanderlust year under his belt, Barton checked into graduate school at Columbia as his mother ordered, but his imagination never left the ocean. He scoured the college library for books on undersea exploration and discovered that people had been ducking themselves underwater using buckets and air chambers for thousands of years. The crush of water pressure and their need to breathe, however, kept them within a few feet of the surface, as he had discovered from his own experiments in Cotuit harbor. In 1690, Edmund Halley, better known for his passion for comets, broke the air and pressure barriers by inventing a weighted wooden trapezoidal box with a glass top in which he could descend for a few minutes to about sixty feet. He also invented an underwater pulley system for delivering air to the diver in separate barrels, but the device still could not transport a man beyond the sunlit shallows near shore. Powerful air pumps, full diving suits that protected a helmeted diver down to about three hundred feet, and new techniques to prevent the bends were just being tested in the mid-twenties, and navy submarines—from which no view outside was possible—had descended to 365 feet. But the abyss remained as unknown and mysterious as outer space.

As Otis Barton turned onto Madison Avenue on Thanksgiving morning in 1926, he glanced back down East Sixty-seventh Street to watch the sun transform the top branches of the bare trees of Central Park into golden lattices against the chrome-blue sky. New York was an exhilarating feast of beauty, but above all it was paradise for a dreamer. Another young man living there that fall, F. Scott Fitzgerald, had declared that the city was leading America on the greatest, gaudiest spree in history as the economic boom reached mythic proportions. Battalions of men in suits scrambled to keep the money machine moving toward their own dreams of permanent prosperity in a market that never went down. Bank chairmen and shoeshine boys shared stock tips, President Calvin Coolidge declared that the business of America was business, and a well-known billionaire told a reporter from *Ladies' Home Journal* that everyone ought to be rich. Otis Barton parked his own money with a conservative investment company and forgot about it except for the checks that arrived every month with ever-increasing amounts on them.

The exhilaration of high times suffused every side of life in New York City. More than 250 plays, musicals, and revues premiered in 1926, including Florenz Ziegfeld's new edition of his Follies called *No Foolin'*, which opened at the Golden Age Theater on Broadway after throngs stood in the street for days to buy tickets. Uptown in Harlem, the silky howl of the Jazz Age poured from hundreds of clubs and out into the city, blending with the tunes of the Gershwins and Irving Berlin. And the miracle of radio had blossomed, so the music also flew through the air and into parlors, kitchens, and bedrooms around America. Two Sundays before Thanksgiving, David Sarnoff had thrown a switch at 8:00 p.m. to broadcast an evening of entertainment over the first radio network, which linked the studios of the National Broadcasting Company with stations in twenty-one cities. The broadcast included the music of the Metropolitan Opera and the New York Philharmonic, and the comedy routine of a hit vaudeville act, Weber and Fields. Millions of people on the Eastern Seaboard tuned in and welcomed New York itself into their homes. Sarnoff's radio network, along with hundreds of magazines and seventeen daily newspapers, were transforming New York as much as the frenzy on Wall Street.

The city, as one of the writers for the two-year-old *New Yorker* quipped, had become a gymnasium of celebrities. Not only could anybody be rich, but anybody could be famous. Actors, journalists, politicians, policemen, gangsters, murderers, nightclub owners, writers, daredevils, baseball players, and even scientists were becoming celebrities. It didn't take much: a group of young critics and writers, including George S. Kaufman, Robert Benchley, and Dorothy Parker, became legendary just for their lunchtime conversations at the Algonquin Hotel. "The publicity machine," observed columnist Walter Lippmann, "scanned the horizon constantly for the event which may become the next nine days' wonder . . . it is like the beam of a powerful lantern which plays capriciously upon the course of events throwing now this and now that into brilliant relief but leaving the rest in comparative darkness."

Otis Barton had always dreamed of adventure, but his current fantasy of exploring the oceanic abyss was also about becoming famous. He loved the vision of himself as a celebrated explorer featured in newspaper stories. While most other men and women dreaming of becoming rich or famous were thinking about the stock market, the stage, or the sensational new world of radio, Barton couldn't take his mind off the deep ocean. And now he had a plan. After the diving experiments of his Cotuit summers and years of engineering classes, he was able to transform elements he had seen in drawings of Alexander the Great in a barrel on the bottom of the sea, Greek underwater saboteurs, and Halley's diving bell into clear pictures of a craft in which he could descend into the depths. Barton's diving machine was a steel sphere at the center of a frame holding chambers of gasoline for controlling buoyancy, with racks of lead balls for ballast that could be released to ascend. The craft would move freely in the water like a submarine, with no cable or air hose. He would get oxygen from tanks, and soda lime and calcium chloride would absorb the products of respiration, carbon dioxide and moisture. But that was as far as he had gotten. He would need a viewing port of some kind or there wasn't much point in going down, and he didn't think ordinary glass was strong enough. He'd need a light, too, but he had no idea how to build one that could withstand the pressure.

Barton was thinking about glass portholes in submarines that Thanksgiving morning when he returned to reality for a moment and handed a newsboy a nickel for the *New York Times*. He glanced at the weather box in the corner—fair and slightly colder today; tomorrow cloudy and warmer—and scanned down to the headlines over the story of a double murder trial at which an astronomer had refuted the testimony of another witness that there was a full moon on the night of the crime. A second headline reported that two seaplanes had crashed en route from the United States to Colón at the northern end of the Panama Canal, and a third that a deal had been sealed to unify the subway systems under a company owned by the city of New York. As Barton took his first steps away from the newsstand, he snapped the paper open for a quick look at the rest of the headlines and froze:

BEEBE TO EXPLORE
OCEAN BED IN TANK
Steel Cylinder Will Withstand
Water's Pressure at Depth
of Mile or More

FOR USE ON EQUATOR TRIP
He will Experiment With It Next
Spring After His Return From
Shark Studies in Haiti

Dr. William Beebe, scientist and explorer, said yesterday that one of America's leading steel corporations was building for him a deep-sea diving tank for exploring ocean bottoms at a depth of a mile or more.

This tank or cylinder will be about a foot and a half in diameter and seven feet high. The steel walls will be about a quarter of an inch thick to withstand the terrific pressure of the water at great depths. It will have a window about 7 by 12 inches in size, made of thick glass capable of withstanding a pressure of several tons to the square inch.

It won't work, Barton thought. The steel of a tank of that shape has to be much thicker to withstand the pressure, so thick that it will be impossibly heavy. Either that or it has to be braced so extensively inside that there won't be room for a passenger. The story concluded with Beebe's optimistic summary of his plans for deep diving. Barton was distraught. William Beebe was one of his idols, an adventurer and author whose popular books about his jungle expeditions had fueled some of Barton's childhood fantasies. Beebe had startled his fans the year before when he announced that he was leaving the jungle to study the deep ocean, and he was already publishing sensational accounts of exploring shallow water in a copper diving helmet. Now the legendary naturalist was confidently promising the world that he would be the first human to descend into the abyss. By the time Barton reached his apartment, his vision of a glamorous future as a bon vivant and celebrated explorer of the ocean depths had dissolved into disappointment and anxiety.

Two

BEEBE

*I want to see you game, boys, I want to see you brave and manly,
and I also want to see you gentle and tender. Be practical as well as
generous in your ideals. Keep your eyes on the stars and keep your feet
on the ground. Courage, hard work, self-mastery, and intelligent effort
are all essential to successful life. Character, in the long run, is the
decisive factor in the life of an individual and of nations alike.*

Theodore Roosevelt, "Youth,"
an inscription in the Rotunda of the
American Museum of Natural History

On the same Thanksgiving morning that Otis Barton walked home nursing his frustrated dream, William Beebe sat reading his own copy of the *Times* in an apartment directly across Central Park at One West Sixty-seventh Street. The newspaper had been delivered to the door of his penthouse in the Hotel des Artistes, a building erected just after the turn of the century as part of the city's plan to draw artists and writers to affordable flats and studios on the Upper West Side. By 1926, though, the neighborhood had become fashionable and expensive, blessed by the park nearby and subways that tied the city into a manageable package no matter where one lived. Beebe was sitting in a Morris chair in his two-story living room lit by floor-to-ceiling factory windows facing south to give a view of the park and, beyond that, the mansard roof of the Plaza Hotel and the turrets of midtown Manhattan. Two walls of the room were lined with books, many of which Beebe had written, and another held a massive fireplace. In a nook, a globe rested in a mahogany cradle under a large model of the steam yacht *Arcturus,* aboard which he had led two expeditions.

Beebe's morning had already included sorting through his notes for a lecture on his recent oceanographic work in the Hudson Canyon off New York Harbor. He was also working on the details of an expedition scheduled for the following year to helmet-dive and drag nets through a particular patch of the Caribbean off Haiti. Most oceanographers were trying to capture and identify as many creatures as possible in large areas of the ocean, but no one was concentrating on understanding the lives of animals in particular small parts of it. On jungle expeditions, Beebe had become convinced that the traditional field practice of collecting animals and ignoring the rest of the neighborhood in which they lived left impor-

tant aspects of the stories of their lives untold. So he came up with the idea of collecting and identifying every animal in a quarter square mile of jungle and trying to figure out how their lives depended upon one another. Now he wanted to do the same thing in the ocean.

The week after Thanksgiving, Beebe was embarking on a three-month lecture tour "to inspire enthusiasm in those whose eyes are just opening to the wild beauties of God's out-of-doors." But at church socials, meetings of literary societies, and events sponsored by the Explorers Club, he was also raising money. Every lecture he gave, every book he wrote, and every mention in the newspapers helped him to promote his work. There was plenty of money around since the stock market was roaring along, but Beebe ran a big operation. The New York Zoological Society had been Beebe's home since Henry Fairfield Osborn, the president of the young society, had given him a job as an assistant curator of ornithology in 1899. But he hadn't remained an assistant curator for long. As his fame grew, the patrons and directors of the Zoological Society realized they had a thoroughbred fund-raiser in their stable, and they gave him rein to explore wherever his curiosity took him. Beebe mounted expeditions, survived attacks with blowguns, spears, and crossbows, and brought back sensational and scientifically valuable accounts of the flora and fauna of Nepal, India, China, Egypt, South America, and dozens of other places. These made great reading in professional journals, books, magazines, and newspapers.

Beebe himself had created the Department of Tropical Research, the only institution of its kind within the Zoological Society, and he had to come up with a lot of cash to keep it going. In 1926 he maintained a full-time staff of research associates, including his protégé and chief assistant, John Tee-Van, who took care of most of the details of running a laboratory at the Bronx Zoo and organizing the department's many expeditions. Tee-Van had come to the zoo with no formal education in science, but Beebe liked him and gave him a set of natural history encyclopedias to see if he could learn. Tee-Van devoured the books and quickly became competent in collecting, taxonomy, and the other essentials of fieldwork. He was as loyal to his director as Sancho Panza to Don Quixote. Beebe's staff and his expeditions to the Sargasso Sea, the Galápagos Islands, and the jungles of South America cost a bundle, and every penny had to come from private donors because governments didn't give money to naturalists.

Beebe didn't particularly like lecturing, but he hit the road every year and lit up rooms with energetic talks that inspired floods of fan letters, donations from schoolchildren, and admission to the inner circles of New York society, where the real money was. He had a gift for putting people in an auditorium or at a dinner table at ease with his curiosity. But he also had a harsh underside and was quite willing to abruptly cut off a friend, colleague, or employee whom he judged to be dishonest, lazy, or disloyal. Beebe couldn't stand mediocrity; he had no use for people who didn't live heroically, and despised the emerging American class of people who believed that a well-lived life consisted of wage earning, too many children, and a bizarre period of idleness before death called retirement. Beebe came from middle-class roots, but he had spent his whole life trying to leave them behind.

He admitted a select few people to what he called his "Crowd," bought them presents, and gave them parties. In New York, Beebe's A-list included politicians, financiers, and artists. Theodore Roosevelt admired Beebe's early work and became a close friend and mentor to the young naturalist; Beebe thought Roosevelt was the greatest American who had ever lived. Just before Roosevelt died in 1919, he wrote what is believed to be his last personal letter to congratulate Beebe on the completion of his manuscript for a definitive book on pheasants. William Harrison, Mortimer Schiff, and other tycoons came to Beebe's costume parties, loaned him their yachts for expeditions, and contributed big money to the Zoological Society specifically for his use. Beebe's Crowd also included Fannie Hurst, a celebrated author and champion of adventurous marriages who lived downstairs from him at the Hotel des Artistes; Rube Goldberg, a syndicated cartoonist noted for his intricate, whimsical drawings of fantastic contraptions; and Zarh Pritchard, a bohemian eccentric who loved the sea and painted gaudy underwater scenes. Beebe loved to dance, enjoyed a drink or two every evening—his favorite cocktail was the brandy Alexander—and was very aware of himself as a dashing figure at the height of the Roaring Twenties.

Beebe was not classically handsome but he was attractive in all the ways that signal power, virility, and energy. He was six feet tall, wiry, bald, fit, quick-witted, charismatic, and quietly dominant in every social and professional situation. Much of his charm came from his boyishness and a vulnerable enthusiasm for everything that interested him, from building a kite to throwing a costume party to mounting a six-month expedition to South America. He was basically a grown-up Boy Scout: trustworthy, loyal, friendly, and courteous, though not always kind.

Beebe was well known both as a literary figure and as a naturalist. By 1926 he had published eleven books and hundreds of popular magazine articles based on his adventures. Tens of thousands of fans around the English-speaking world anxiously awaited every word he wrote. He loved storytelling, and an evening with Beebe at home or in a jungle tent usually featured him reading some Rudyard Kipling, Jules Verne, H. G. Wells, or Lewis Carroll. When he traveled, he brought along a small library, which included *Alice in Wonderland,* his favorite book; *Introduction to the History of Science,* by George Sarton; *The Origin and Evolution of Life,* by Henry F. Osborn; *Science and the Modern World,* by A. N. Whitehead; *The Depths of the Ocean: A General Account of the Modern Science of Oceanography Based Largely on the Scientific Researches of the Steamer Michael Sars in the North Atlantic,* by John Murray and Johan Hjort; and A. A. Milne's *Winnie the Pooh.* Beebe wrote compulsively and quickly, churning out his books and articles from detailed notes and keeping a daily diary since he was thirteen.

On page one of his first diary, he wrote the words *Naturalist's Diary* in large letters of flowery script over his inaugural entry.

Aug. 16, 1890
Saw two or three P. Troishus, P. Turuus and P. Asterial also an Archippus and a N. Ephestion (Loc.) Llewellen Park. Collected spec. P. Asterial, P. Lucia, Wasp, Weavil and G. Philachie.

On the inside cover facing the first page of his *Naturalist's Diary* he drew a map of his neighborhood in East Orange, New Jersey, a rural village of about ten thousand to which his parents had moved from Brooklyn just after Will started grade school. Beebe's father had been born in Brooklyn, spent his working life as a clerk in his own father's paper business with offices on Nassau Street in Manhattan, and was most proud of his service as an officer in the Brooklyn Gatling Gun Battery of the New York National Guard. His mother, Henrietta Marie Younglove, was a woman of intense drive, a bluestocking Christian Scientist from Glens Falls whose main interest, other than her church, was seeing that her only child achieved his potential. (Their second son, John, had been born in Brooklyn when Will was four years old but died fifteen months later.) In East Orange, the Beebes lived in two rented houses and eventually settled permanently in a big wood-frame Victorian at 73 Ashland Avenue with a full third story lighted by five dormers and decorated with wooden scrollwork. With its maple and chestnut trees and a backyard with a garden and carriage house, it was exactly what Charles Beebe had in mind for living happily in safe, bucolic surroundings with his wife and son.

By the time Will was a teenager, the family was financially well off and able to take summer vacations in the hills of Pennsylvania, Nova Scotia, and New Brunswick, during which his passion for the natural world took wing. He loved camping, hiking, and most of all collecting, identifying, and preserving birds, insects, and anything else that inhabited the mysterious world of the wild. At home in New Jersey, he was preoccupied with his investigation of the flora and fauna of Orange Mountain, a forested ridge about two miles from Ashland Avenue. His favorite book when he was a teenager was *The American Boy's Handy Book* by Daniel Carter Beard, one of the founders of the Boy Scouts of America, and Will tried everything it suggested. From it, he learned to build kites, aquariums, boats, and snow houses, use a bird call, camp without a tent, catch and stuff birds, fish, train dogs, and stage puppet shows.

In his dairy, Beebe painstakingly recorded daily events and encounters with animals.

My expectations for this year of 1894 are as follows:

In New Jersey, I expect to see at least 60 kinds of birds, 6 kinds of mammals, 6 kinds of reptiles, 6 kinds of fishes.

I expect to kill at least 100 animals and birds, say 25 varieties.
I expect to collect 100 insects (50 var.).
I expect to make notes on at least 250 days in the year.
I expect to see at least 4000 crows migrating.

In Pennsylvania
I expect to see at least 50 kinds of birds, 6 kinds of mammals
I expect to kill at least 50 animals and birds, say 25 varieties
I expect to collect at least 300 insects (75 var.)

When his mother enrolled him at East Orange High School, there was no doubt in her mind that her son was a scientist, and Henrietta Beebe encouraged her son's passion for nature like a stage mother clucking around a promising young dancer. She guided his curriculum, insisting that he take four years of Latin, two of German, four years of English and rhetoric, and a series of six courses in physics, chemistry, geology, botany, physiology, and zoology. He was a B-average student, but his enthusiasm, curiosity, and endless energy placed him first in the minds of his teachers in terms of his potential.

When Will graduated, his mother took him to New York, convinced paleontologist Henry Fairfield Osborn and ornithologist Frank Chapman at the American Museum of Natural History that her son was a prodigy, and handed his education over to two of the reigning giants of natural history. They advised him to enroll as a special student in the Department of Zoology at Columbia, where Beebe spent three years going to lectures by his two mentors as well as anthropologist Franz Boas, evolutionary biologist Keith Brooks, and other luminaries who were drawn to the New York campus because it was the center of American inquiry into the great puzzles of the natural world. He took only a few formal courses, though, and in the fall of 1899 Beebe left Columbia without applying for a degree to take a job as an assistant curator of ornithology at the New York Zoological Park (the Bronx Zoo).

Beebe thought that the zoo was the best place in the world for a man like him to work, but he had no idea that he would never leave. He reported for work on October 16, 1899, and drew his first pay on November 4, six five-dollar bills that looked very large to him. With his first earnings he paid his room and board, bought his mother a picture of the biblical prophets for $3.75, his father a smoking jacket for $2.45, and spent the rest on a subscription to a literary digest, a frame for an Audubon print, and some photograph albums.

Young Beebe lived prudently, carefully rooting himself in reality, obligation, and the wonders of the natural world, but in his last diary entry of the 1800s, the fully fledged William Beebe gave himself mixed reviews.

Sunday, Dec. 31, 1899
The last day of the Century! and how little that seems to convey. I have read a lot about the "New Thought" today in my office at Bronx Park and wish I could understand it more. A new century! No more 18 hundreds. It

*makes me very sad to think of the change. Tomorrow in every thing else
will be the same as today but how great the gulf the little figures seem to
make. The papers and magazines are filled with comparisons of this century with former ones, but all this reading makes less impression on my
mind than my own few thoughts.*

*It is new for me in a great many ways—a new job; a new home; new
opportunities; new friends; more money than I ever had before, and yet I
am not happy altogether. But I am so unsettled in my outside lines of work
that I hope soon to spend each day more profitably and* better.

*This is an epoch in my 22nd year which will never come again in my
lifetime. Tomorrow—Monday—1900 a new year and* Century. *What a
day for new resolves to be made and kept.*

*I have hundreds that cannot be written but although I will probably
back them many times, I will keep them in the end.*

*December 31st 1899
I have spoken!*

On Thanksgiving morning twenty-six years later, Beebe got around to
taking a quick look through the *New York Times* to find a story based on
an interview he gave to one of its reporters the day before. He dropped
his eyes right to the headline containing his name and noted happily that
it was above the fold, next to the most important stories of the day. He
winced when he read the first sentence, though, because he was not *Dr.*
Beebe at all. He had never even received a bachelor's degree, let alone a
doctorate. But the truth was that if the public thought of him as Dr.
Beebe or Professor Beebe or Director Beebe, it certainly didn't hurt when
it came to fund-raising.

Beebe was confident, though, that his education, experience, and list
of publications were worthy of whatever honorific people attached to his
name. He had evolved from a thirteen-year-old boy making notes about
birds in his yard into a larger, older version of that same self just living in
a bigger neighborhood. Beebe published his first story, "The Bird Called
Brown Creeper," in *Harper's Young People* in January 1895, when he was
eighteen years old. It was a 400-word article on the habits of an incon-
spicuous little bird in New Jersey. His editor dubbed young Will "An
Observing Knight of Harper's Round Table." After Beebe's debut, he

wrote for *Bird Lore,* the *New York Evening Mail and Express,* and the *New York Evening Post;* finally, in 1901, his article on the sandhill crane was accepted by a recognized scientific journal, the *Bulletin of the New York Zoological Society.* He climbed to the top ranks of popular writers because he was willing to rhapsodize about nature as well as study it, and this appealed to legions of lay readers. He had an instinctive feel for metaphor, simile, and analogy, and a knack for transforming what he saw in the field into familiar imagery. After he went helmet diving in the Sargasso Sea, for example, he made a note in his diary that from beneath the surface the floating weed reminded him of a grape arbor, and then passed that image to readers in his latest book, *Arcturus Adventures,* along with the fact that the word "sargasso" comes from the Portuguese *salgazo,* which means "little grapes." Beebe knew that the astonishing, undulating sea of grass that dominates the Atlantic for thousands of square miles in a great triangle south of Bermuda is much easier to visualize as a giant vineyard for a reader in an armchair.

To a practicing botanist or biologist of the time, though, Beebe's willingness to filter the Sargasso Sea through the lenses of literary devices and technique made him suspect. "Real" scientists did not pander to general audiences, report the findings of their fieldwork in best sellers, or make promises and plans on the front pages of the daily newspapers. And what's more, "real" scientists did not get divorced with messy publicity as had Beebe, or go dancing in jazz clubs, or bring attractive young women with them on expeditions. Beebe had published dozens of papers in the most reputable scientific journals—*Science, Auk, American Naturalist,* and *Zoologica*—but still his critics chortled every time a story appeared in the *Ladies' Home Journal, Boys and Girls,* or even the *Atlantic Monthly.*

William Beebe was balanced successfully on a tightrope strung between the edifices of celebrity and science, but he worried constantly about falling off. He considered himself to be eminently responsible as a scientist and could not stand being dismissed from the ranks of that profession. Nor could he abdicate his throne as America's most popular natural history author, especially now that he needed the money for ocean exploration, which was proving to be more expensive than birds or jungles. To most scientists, abandoning ornithology in midcareer to study the sea and its creatures would have been unthinkable, but that's just what Beebe had done in the spring of 1925 after making a helmet dive in the Galápagos.

The oceans cover 71 percent of the surface of the earth with 328 million cubic miles or 361,200,000,000,000,000,000 gallons of water. Oxygen and hydrogen make up 96.5 percent of ocean water; the other 3.5 percent is dissolved elements, such as chlorine, sodium, and other salts. The average salt content of seawater measures 34.7 parts per thousand, roughly the same proportion as human blood. Most of the salt probably leached or blew into the sea from land. Neither Beebe nor anyone else at the time knew the origin of the ocean, its greatest depths, the contours of even a fraction of its floor, or the nature of life under pressure in total darkness. After his dive in the Galápagos, though, Beebe did know that the exploration of the ocean was the greatest scientific adventure of his lifetime, and he wasn't about to miss it.

The first systematic study of the world's oceans was not attempted until 1872, when HMS *Challenger* and her crew left England on a four-year cruise around the world. She was a 226-foot Royal Navy auxiliary corvette from which all but two of her guns had been removed to make room for specimen tanks, nets, cables, and crates of preservatives, books, and dissecting instruments. The *Challenger* sailed with a crew of 243 men, a third of whom were scientists and their assistants, easily the most unusual complement ever to sail on a British man-of-war. C. Wyville Thomson, a scientist who had organized earlier, more modest voyages, led the ambitious expedition. He wanted to map the seafloor to find out if it was possible to lay a telegraph cable across the ocean between Europe and America. He also wanted to sample the temperatures of different layers of seawater with a newly invented instrument called a reversing thermometer that could be triggered to operate and store its readings. The prevailing notion, called the azoic theory, was that the deep ocean was uniformly cold and could not sustain life. "The oceans are wastes of utter darkness," Thomson had said before his first ocean expeditions, "subject to such stupendous pressure as to make life of any kind impossible." After his earlier voyages, though, he had seen such an astonishing diversity of life from the depths of the ocean he was well on his way to admitting that he might have been wrong. On the *Challenger* he hoped to prove that there were layers and variations of temperature, and continue to dismantle the azoic theory that he had once supported.

Thomson and his colleagues were also anxious to prove or disprove another theory that was adding a complex new dimension to the study of life on earth. Just a few years earlier, in 1859, Charles Darwin had described a natural process by which entire groups of animals became extinct and new ones evolved over hundreds of thousands, maybe even millions of years. Darwin had built his revolutionary theory on his observations of new kinds of animals that seemed to have evolved because of climate shifts, geologic separation, or other environmental upheaval which separated populations and forced them to become different over time. Darwin also hypothesized that if trilobites, ammonites, sea scorpions, and other animals lived in a relatively unchanging environment such as the deep ocean, they might still be there. On his earlier voyages aboard the smaller HMS *Porcupine,* HMS *Shearwater,* and HMS *Lightning* with dredges that could reach to depths of almost a mile, Thomson had dragged up some bizarre creatures, but none of the prehistoric ones he'd hoped to find. Perhaps, he and his many supporters theorized, the living fossils were simply in deeper water.

The *Challenger* carried rope and cable with which her crew could sound the depths of the ocean to over five miles. They discovered that the continents were girdled by underwater shelves that gave way to deep plateaus, and that the seafloor was fractured by high mountain ranges and trenches much deeper than their probes could reach. The ship also carried an eighteen-horsepower steam winch that allowed them to dredge the sea bottom. The expedition collected, dead and alive, 4,717 creatures no one had ever seen before. They found none of the living fossils they hoped were there, but the expedition put an end to the notion that the depths of the ocean were "wastes of utter darkness." Dozens of expeditions followed the *Challenger,* permanent research stations bloomed around the world, and the science of oceanography was transforming the study of life. But no one had descended into the abyss to actually witness the wonders of the hidden new world.

For the better part of two years, William Beebe had been patiently towing nets through deep water and bringing up animals he identified from drawings and descriptions made by other scientists. Some of the creatures he landed, though, had never been seen by anyone before him. In his lab aboard the *Arcturus,* he had sealed out sunlight and marveled at the blazing galaxies in his specimen trays and tanks: some alive, most dead, but all creatures from a universe clearly unlike that of animals found on the surface. In relationship to the creatures of the deep, Beebe compared himself to a student of African zoology who has trapped a few rats and mice, but is completely unaware of antelope, elephants, lions, and rhinos. Beebe's imagination was overwhelmed with visions of what might really live in the abyss.

Still, leaving the jungle for the ocean presented enormous risks to his reputation, and the front page of the *Times* reminded Beebe that New York City was an awfully public place in which to do it.

Three

A DAY AT THE ZOO

*Members of the scientific staff of the Park and of the Aquarium did
not, however, enter the well trodden field of the lifeless cabinet or
museum animal, nor of the older systematic or descriptive zoology,
but sought a new and inspiring field which had been relatively
little pursued, namely, the observation of the living bird and the
living mammal, wherever possible in its own environment.*

Henry Fairfield Osborn,
Preface to *Zoologica,* Volume 1

For months after Otis Barton read the devastating news that William
Beebe was launching his own campaign to explore the deep ocean,
he continued to put one foot in front of the other; but without his dream
of diving into the abyss, he was aimless. His lackluster work at Columbia
University only reminded him of what he was already beginning to think
of as his wasted, insignificant life. He still missed his younger sister, Ellen,
who had died suddenly four years earlier, and the bleakness of mourning
added another dark shade to his depression. Barton went to a few classes,
walked across the park to Broadway for movies, and took occasional trips
to Boston to see his mother, Katy Gaule, Francis, and Mary; but he was
miserable.

Finally, Barton's imagination and youthful optimism rescued him: he
constructed another grand fantasy in which he could live happily. Barton
decided that if Beebe was going to be the first man to explore the depths
of the ocean, he would go fossil hunting like paleontologist Roy Chapman Andrews. On his field trips to Outer Mongolia, Andrews played
the role of a swashbuckling naturalist to the hilt, wearing a ranger hat,
prominently carrying a revolver at his hip, and telling people that he
honestly believed he was born to be an explorer. In his teens, Andrews

had happily taken a job scrubbing the floors of the American Museum of Natural History to get his foot in the door. He quickly became a member of the collecting staff and the expert on whales and other large mammals. By 1920 Andrews had developed an obsession with his pet hypothesis that the fossils of the Central Asian plateau contained evidence that the region was the source point from which all mammals had radiated around the world. With the blessing of the museum's president, Henry Fairfield Osborn, he set about trying to prove his theory on several expeditions to reconstruct the history of the plateau and its fauna, geology, climate, and vegetation. It was an enormous undertaking, but typical of a grandiose era when the earth and the history of its inhabitants were being systematically measured for the first time. For two years, Andrews searched the Gobi Desert with little success, but he eventually found the fossilized bones of a giant beast that had lived during the last ice age. He battled bandits to get them back to New York and dazzled Osborn and the rest of the museum scientists with his finds, though his theory of a Central Asian source point unraveled after much more defensible discoveries were made in Africa. Roy Chapman Andrews fit perfectly into Otis Barton's new vision of himself.

In the summer of 1927, Barton went on a geology field trip to Wyoming with a fellow Columbia graduate student, Eugene Callihan, whom Barton described as "a tall, rangy Oregon boy who could out-walk, out-dig and out-collect the rest of us, was no scatter-brained day dreamer like me, no hero worshipper. He didn't see himself surrounded by an aura of glory as a junior Beebe or Andrews but he was determined to become a first-rate geologist." Barton proposed that the two of them spend the summer of the next year, 1928, in Persia, exploring the western edge of the Central Asian plateau, perhaps adding to Andrews's work and to their own reputations. Like everyone else with capital to invest in the stock market, Barton was getting richer by the hour; so he would pay the bills, among them the cost of special expedition letterhead:

WESTERN ASIATIC EXPEDITION

OTIS BARTON, ZOOLOGIST

EUGENE CALLIHAN, GEOLOGIST

During the months between Barton's conception of the Western Asiatic Expedition in the summer of 1927 and his departure in the spring of 1928, he kept tabs on William Beebe with the hope that things would not go well with his plans for exploring the deep ocean. He read the daily papers and kept up with reports of the Department of Tropical Research at the Zoological Society, and it looked as though Beebe was not making much progress with his diving tank. There were more stories in which Beebe insisted that he was going to explore the deep ocean, but in the newspaper illustrations, his contraption still looked to Barton like a laundry boiler, and he remained skeptical about its ability to withstand the enormous pressure. To Barton, it was obvious that a sphere was the strongest shape for a diving tank. He knew Beebe would have consulted an engineer before announcing his plans, but he also knew that no engineer had ever designed a successful deep-diving craft, so obviously mistakes were possible.

Barton read Beebe's accounts of helmet diving off Haiti and trawling in the Hudson Canyon off New York, a few of his stories on birds, and the announcement on the society pages that Beebe had gotten married for the second time in the summer of 1927. Every member of New York's science and literary establishments, including Otis Barton, was aware of Beebe's marital history and his reputation as a man to whom women were drawn like moths to a field lantern. Beebe had married his first wife, Mary Rice Blair, in August 1902, when he was twenty-five years old, and the following year they headed south on a belated honeymoon to identify as many Mexican birds as possible. In 1905, Beebe published *Two Bird Lovers in Mexico,* an account of that journey and the debut of the compelling blend of romance and scientific inquiry that characterized his books from then on and eventually accounted for his immense popularity. More trips with Mary followed, including a stupendous 52,000-mile odyssey to survey the world's pheasants. The couple sailed from New York aboard the *Lusitania,* and their itinerary was the envy of many other young adventurers: Ceylon, Calcutta, Darjeeling, Tibet, Burma, Singapore, Malaya, Yunan, Mongolia, Japan, and back to America. They dodged cholera on the subcontinent, risked plague in China, suffered constantly from seasickness when offshore, but returned with the world's most complete list of pheasants. That expedition led to the 1922 publication of Beebe's magnum opus, *A Monograph of Pheasants,* a book that stunned the ornithological world with its immensity, passion, and beautiful illustrations, and solidified his reputation as a naturalist and author.

Beebe and Blair returned from New York from the Pheasant odyssey in May, 1911, and on January 13, 1913, she left him and moved back to Virginia to live with her parents. Two months later she took the train to Reno, served the brief residency prescribed in those days, and was granted a divorce on the grounds of "cruel and abusive treatment." In the court filings in Nevada, Blair alleged that Beebe had subjected her to financial stringency coupled with reckless spending, unexpected guests, nasty disputes with neighbors, verbal abuse, long silences, unexplained absences, and suicide threats. According to Blair, who probably resorted to hyperbole to build a case for a divorce, Beebe had tried to take the upper hand in screaming arguments by placing a pistol in his mouth and threatening to throw himself into a jungle river or cut his own throat with a razor. A one-page letter from Beebe's lawyer denied the allegations but offered no explanations or proof to the contrary, and he did not appear in the Nevada courtroom when Judge Cole L. Harwood granted the divorce on August 29, 1913. The headline in the *New York Times* the next day was "Naturalist Was Cruel." Mary Blair moved back to New York and promptly married Robin Niles, Beebe's friend, neighbor, and the son of one of the founders of the New York Zoological Society. Beebe was crushed, and he excised any mention of Mary and her immense contributions to his work from the manuscript of *A Monograph of Pheasants.* Her literary banishment, along with the divorce itself, added a controversial dimension to his celebrity, which painted him as either a misogynist or a self-serving egotist determined to share the limelight with no one.

Shaken by Mary's bitter departure and her humiliating marriage to Niles, Beebe went to war in 1917 as a member of the French Aviation Service, trained as a pilot, but never flew in combat. Even in the brutal environs of the battlefield he remained a naturalist. He returned to New York, slightly wounded on the wrist from a training accident, with enough systematic inquiry recorded in his notebooks to write and publish a bizarre article, "Animal Life at the Front," and his first science adventure novel, *Jungle Peace,* in which he embedded his philosophy of the healing power of nature.

> After creeping through slime-filled holes beneath the shrieking of swift metal, after splashing one's plane through companionable clouds three miles above the little jagged, hero-filled ditches, and dodging other sudden-born clouds of nauseous fumes and blasting heart of steel; after these, one craves thoughts of comfortable hens, sweet apple orchards, or ineffable themes of opera. And when nerves have cried for a time "enough" and an unsteady hand threatens to turn a joy stick into a sign post to Charon, the mind seeks amelioration—some symbol of worthy content and peace—and for my part, I turn with all desire to the jungles of the tropics.

After Mary Blair and the war, Beebe was resigned to life as a bachelor, but in 1927, he was smitten by a young writer from Nebraska named Elswyth Thane Ricker. Thane, who never used her real last name and in fact concealed it, was svelte and stylish with coils of golden brown hair, the commanding presence of a woman much older than her twenty-three years, and a self-confidence that instantly appealed to the fifty-year-old Beebe. Elswyth Thane told him that she had idolized him since childhood, devoured all his books, and hoped he didn't mind that she had dedicated her first novel to him. *Riders of the Wind* is a bodice ripper set on the Northwest Frontier of India in which a tall, lean, intrepid hero and a considerably younger heroine fall in love.

So, while the cold dawn gained on them slowly, they talked as they had never before talked, no longer shy of things, no longer tacitly avoiding things, their quaint, school-boyish reticence laid aside for this last great moment that remained to them, and there were many things that neither of them could find words for, but these were all encompassed in their boundless understanding, each for the other, which filled in the gaps and replied to them.

Beebe was quite comfortable being adored and he was bowled over by Thane's youthful beauty; she was not at all possessive, and they quickly learned that their shared lives as writers helped them sympathize with each other's busy, solitary days. On September 22, 1927, Beebe and Thane were married aboard Harrison Williams's yacht *Warrior*, anchored in Oyster Bay, Long Island, with a full roster of high-society guests.

In the year after his marriage to Elswyth Thane, Beebe's output of books and magazine articles fell to a trickle, and by March of 1928, Otis Barton figured that Beebe might be far enough off the track of a workable diving tank that he still had a chance. Collecting bones in Asia couldn't compare with becoming the first explorer to descend into the abyss, so Barton hired a marine architect to translate his rough sketches of a deep-diving craft into a finished engineering design. He asked around among his Park Avenue friends and learned that the firm of Cox & Stevens was the maritime equivalent of a couture designer, with the owners of some of the most elegant and expensive yachts in the world on its client list. Barton called Irving Cox, and then wrote:

I have had a conversation with Mr. Cox and am acting in accordance with his instruction in forwarding to you the first blue prints of this deep sea tank. I am anxious to have more preliminary drawings of this tank. It is to be used to make submarine observations at great depths for purely scientific purposes. I should like very much to have two sets of drawings, one for a tank ca[pa]ble of withstanding 10,000 ft. of water, and another for 5,000 feet. I also wish some idea of the cost of constructing each.

If you will be kind enough to make these plans for me, and also a price estimate, I will then be able to present them to scientists interested in the project. I should like to have the preliminary drawings within a few weeks if convenient for you.

I enclose the blue print to give an idea of what is wanted and not as a guide to your engineer.

I am a graduate student in science at Columbia and the American Museum of Nat. History. Some funds have actually been raised for this enterprise.

I wish to thank you very much for the interest you have already shown in the matter.

After Barton sent Irving Cox the sketches of his diving tank, he sailed for Persia with Eugene Callihan in early June. They toured Arabia for the summer, at one point stopping long enough to excavate a four-million-year-old streambed in which they found the bones of rhinos, horses, hyenas, pigs, and antelopes, and a single tooth of a Silvatherium, a rare animal related to modern giraffes. In Constantinople, as the July heat and the fading hope of glory in the fossil beds were wearing him down, Barton scribbled a hopeful note on expedition stationery to Cox & Stevens: "I hope very much that you will have made some plans for my tank by the time I get back in September. I expect I shall then be able to take the matter up actively at the American Museum."

Barton went back to New York in the fall of 1928 and took an apartment in a nine-story brick building on East Eighty-second Street between Park and Lexington, directly across Central Park from the American Museum of Natural History. The collection of bones and fossils from the Western Asiatic Expedition was neither significant nor original, but it was enough for Barton to declare himself a success. Through the autumn, he worked on an account of his expedition for the magazine of the museum, *Natural History,* under the title "Fossil Bones in a Persian Garden: Remains of Animals Caught by Streams and Buried in their Deposits Fifteen Million Years Ago Come to Light Amid the Fruits and Flowers of Persia."

"Northern Persia is the greatest carpet market of the world," Barton wrote of his arrival in the country. "Even stables and garages have carpets on their floors, that elsewhere would grace a palace. The streets of the great bazaar at Tabriz are covered with carpets. Here we found our way about with great difficulty, for the mud walls around the houses and gardens made the crooked streets look so much alike that we had difficulty in identifying them."

The article was more about carpets, Barton's encounters with border guards who took his revolver away from him, and his flight over the desert on Imperial Airways than scientific discovery, but it did feature a photograph of him in a pith helmet and a khaki field shirt kneeling over the fossil of a wolf. If nothing else, the Persian adventure boosted Barton's confidence and inspired him to go back to work on his diving tank.

Encouraged by his modicum of success, Barton struck on the idea of simply talking to Beebe and offering to join forces with him as a fellow explorer. To increase the chances that Beebe would agree, Barton sketched out a new version of a craft that combined his own earlier design based on a steel sphere with Beebe's idea of a structure suspended on a cable instead of moving freely in the water. He called Irving Cox, told him to stop work on the earlier design, and made an appointment to discuss his new idea. In early October, Barton walked into a wainscoted corner office on Madison Avenue, where Cox shook his hand, led him to a chart table, and asked the nervous young man to show him what he had in mind. Barton laid out his sketches and told Cox his theory that a sphere was the only shape that would allow a human to survive a deep descent. The infinite arcs of the sphere, Barton said, distribute the crushing pressure of the water evenly across the steel surface. Any other shape with flat surfaces, such as a cylinder, would have to be internally braced so heavily that there wouldn't be room for passengers. The size of the sphere was critical. A solid steel ball could be lowered to great depths and still remain intact. A hollow ball, though, had to be the right diameter and its walls the right thickness to carry a man safely. If the sphere was too big, the volume of air inside would make it too buoyant to sink without extra weight; too small, and there would be no room for observers.

Cox had designed hundreds of yachts and ships of wood, iron, and steel, but nothing like this, and his mind clicked through the pros and cons of risking the reputation of his company by taking the job. Barton sensed that Cox was giving in to skepticism, so he appealed to his sense of adventure. He stretched the truth and told him that William Beebe might join him for the deep-sea dives, because the time had come for man to explore the abyss that was the last great mystery on earth, that the rewards of success were worth every risk to life and reputation. Barton also said that he would pay cash for the design and that his bank in

Boston would verify his credit that day. Whatever convinced him, Cox told Barton that his firm would be honored to take part in the great adventure, though he would not personally handle the work. He told Barton that Captain John H. J. Butler would be entrusted with the actual design and construction of the world's first deep-sea diving craft, assisted by draftsman H. E. Barrett.

John Butler, a big man with full whiskers and a commanding presence, was an old hand at both helm and drafting table, and though he preferred working on yachts, he was fascinated by Barton's diving tank. Like all sailors, Butler had whiled away countless hours at sea wondering what lay beneath the keel of his ship, and here was a man who proposed to find out. He tackled the preliminary design with the same efficiency he brought to building a ship. Though Barton would quickly become a bit of a pest, Butler respected his willingness to spend money on exploration and his credentials from the American Museum of Natural History. Because of Barton's respectable showing in Persia, William Gregory, an ichthyologist at the museum and a professor at Columbia, was allowing him to use the name of the institution in his dealings with Cox & Stevens and the other companies involved in designing the revolutionary diving craft. The museum also promised Barton a place on its property on the Upper West Side of Manhattan to which the sphere could be shipped to be fitted out for its first expedition. Barton wasn't exactly sure what would happen after the sphere got to the museum, but he forged ahead.

On December 18, 1928, Barton received a set of finished blueprints of the design from Cox & Stevens and a letter from John Butler.

Dear Mr. Barton:

Enclosed herewith are two prints of our design on the Deep Sea Tank suitable to descend to a depth of one mile.

You will note that we have decreased the internal diameter to five feet, as per your verbal request of last week.

The housing for the electric lights will be developed in greater detail when the plan is submitted to the foundries for a final price; but the general idea will be as shown on the prints.

We also enclose copies of letters received from the American Steel Foundries regarding the casting, from Roebling regarding the proper steel rope necessary, and from Merritt Chapman regarding the chartering of a suitable tug.

The three cable companies with whom we conferred are unwilling to charter their steamers for more than a day or two as they have only sufficient vessels necessary for their own needs, and require them at all times within their beck and call.

The complete tank and fittings, excluding rope, will weigh approximately seven tons, and with all available figures we estimate the cost of tank to be approximately $5,300.

As we informed you in our letters of December 6th, the steel rope necessary to lower the tank to the required depth should not be less than 2" diameter. The cost of this rope is $1.85 per foot with a discount of from five to twenty percent; and 6,000 feet would cost say $8,900. The drum proper to use in the winding of this rope, would have to have a barrel diameter of seven feet with a 15" depth of flange, and a 62" face. This drum is furnished with the rope.

So we believe a figure of $14,000 to $15,000 will cover the entire construction, excluding the apparatus and other instruments with which you will equip the tank. The figure may possibly be below the above mentioned but we feel that you would rather be on the safe side and have something for which to look forward.

The letter from the American Steel Foundry gives you a fair idea of the problems with which we have had to contend, namely the size and peculiar shape of the casting, being in one piece (the more advisable, we believe) and the immense pressure to which it will be subjected. But we have taken a great interest in the idea and allowed ample factors of safety.

As the rough casting can be furnished in four to six weeks from receipt of order, allowing ample time for machining and fitting up, we believe the tank in its entirety can be completed in from two to three months.

Now, Mr. Barton, we hope that this letter and the enclosures give you all required information, but should any item be overlooked that you wish to have available to take "down East" with you or should you wish additional blueprints, just phone and we will endeavor to fix matters.

Wishing you a Merry Christmas and a Bright New Year, we are

Very truly yours,
COX & STEVENS, Inc.

Through October and November, Barton had written to Beebe every week to ask for an audience, but it was as if his letters had been flying off into space instead of just uptown to the zoo in the Bronx. Barton knew that Beebe was constantly barraged by crackpots and opportunists and was as unapproachable as a movie star, but the more he was ignored the more determined he became to get in to see the great man. In desperation, and with his blueprints in hand, Barton called on a newspaperwoman at the *New York Times* who was a friend of Beebe's, showed her the plans, and begged her to introduce him. Just before Christmas, she called Beebe and arranged an appointment. "You'd better see Otis's blueprints," she told him, "unless you want to lose out on this deep-sea exploration business."

With the letter from Cox & Stevens and the blueprints tucked under his arm, Barton took the subway up to the Bronx on a cold December day in 1928 to meet William Beebe for the first time. The Zoological Park in the Bronx was alive with the aromas, chattering, roaring, and snuffling of hundreds of animals exhibited in the Lion House, Zebra House, Monkey House, Reptile House, and the other ornate brick buildings that form the central quadrangle known as Baird Court. On the outside perimeter to the east, a giant walk-though aviary held hundreds of birds and cast another symphony of sound into the cold air. The aviary was one of the centerpieces of the park, the first of its kind anywhere in the world. It was a pet project of William Beebe, who built it because he couldn't stand to see birds confined to small cages.

The Bronx River runs through the zoo property just to the east of the main buildings, near the end of its short journey from the highlands of Westchester County to the East River and Long Island Sound. The towns upstream were finally treating their waste before pumping it out to sea through the watershed, so the river was recovering from a half century as an open sewer. North of the zoo lay the New York Botanical Garden, an enormous area of well-tended public land that had once grown feed for the city's thousands of horses, which had disappeared almost overnight with the arrival of the automobile. The Bronx was a fashionable part of New York, split in half by the Grand Concourse, a broad, busy boulevard lined with ornate apartment buildings and connected to Manhattan since 1903 by the IRT subway line. The Loew's Paradise Theater was under construction, set to open in 1929, with a planned four thousand seats and a baroque decor that would include a ceiling painted dark blue to resemble the night sky, with small lightbulbs for stars and simulated clouds.

The New York Zoological Society, custodian of the zoo and the aquarium downtown in Battery Park, was founded in 1885. At the peak of the Roaring Twenties it was a very well-heeled institution. In 1928 alone, gifts to the society included a cool million from Anna M. Harkness, and its list of regular donors included retail tycoon Marshall Field, aviation pioneer William Boeing, financiers Mortimer Schiff, Harrison Williams, and Ogden Phipps, and dozens of other prominent New Yorkers. The city itself contributed $50,000. A board of managers ran the society on behalf of its thousands of members, published a journal, and kept the fund-raising machine chugging along, led in 1928 by William Beebe's close friend and champion, Madison Grant. Grant was a lawyer, a former hunting buddy of Theodore Roosevelt, a founder of the Boone and Crockett Club, and a driving force in the creation of the Zoological Society and the park in the Bronx. He had also played a key role in driving Tammany Hall out of New York politics.

Beebe's office at the zoo was splendid, walled with shelves of memorabilia and books, maps, photographs, skulls, mounted birds, and jarred specimens. He did most of his work in his laboratory in another building, so the office had the neat, well-tended feel of a private museum. When Otis Barton arrived for his appointment, Beebe greeted him with a single-pump handshake, took his overcoat, and showed him to a chair. He hung the coat on a rack near the door, took his own chair behind his oak desk, and sized up the young man in a dark blue suit who sat with his hands fidgeting on his lap while his eyes flashed around the room, obviously a bit stunned by the moment. Used to such reactions to his office, Beebe gave Barton a moment to settle down, but since there was no small talk in him, he quickly got to the point. "So," he said, "you've invented a diving device that will work. I won't have any air tubes from the surface. But I suppose you understand that."

Barton had never heard Beebe speak until then, and he was startled by the blunt consonants and warped vowels of Flatbush Avenue coming from the mouth of a man who looked like a speaker of the king's English. He nodded, gestured to his roll of blueprints for permission, stood, and unrolled his plan on the desk. Barton's voice cracked as he blurted his first words since he walked in the door. "It's a sphere." He then ran a finger over the white lines of the blueprint as he pointed out the details of a fourteen-inch hatch, three portholes opposite the hatch, and a lug on top to which a cable could be connected. Dimensional lines registered the inside diameter at five feet, just room enough for two men, Barton said, and Beebe realized for the first time that this design carried the implicit proposal that the young man was planning to join him in the sphere. The walls of the craft would be two inches thick to withstand water pressure down to a depth of ten thousand feet. Barton pointed out an interior cutaway showing the oxygen tanks and chemical trays to remove carbon dioxide and moisture from the air. The details of the windows and a searchlight, Barton admitted, had not been finalized, but John Butler at Cox & Stevens would soon report on the possibilities. A cable strong enough to hold the steel sphere was no problem.

Beebe liked the design. It was simple, and he remembered that years earlier, none other than Theodore Roosevelt had suggested that a sphere with thick enough walls could probably withstand the pressures of the deep ocean. Barton and his architect Butler apparently understood the physics of water pressure better than the men who had recommended a braced cylinder to him two years earlier. On the spot, Beebe and Barton struck a deal. Barton would pay for the sphere, cable, and other equipment, which would cost about $15,000. He would also supervise the construction, leaving Beebe free to continue other work. In return, Barton would own the sphere and accompany Beebe on the expedition, which would be an official venture of the Zoological Society, beginning in the summer of 1929. Beebe would raise the money for all other expenses, including the cost of chartering a big enough boat with a winch to launch and recover the sphere. Beebe told Barton that he already had a good winch, salvaged from his research ship the *Arcturus,* and that he was close to concluding a deal that would give him a full-time field station on Nonsuch Island, Bermuda. It would be the perfect base from which to stage their dives, less than ten miles offshore in water that was more than two miles deep.

Four

SPHERE

B arton was ecstatic as he rode the subway home from the zoo. He called John Butler to tell him that Beebe was in, but he wanted to dive the following summer, so the tank had to be ready in six months. The kid had a lot of sand, Butler had to admit, but he wasn't looking forward to a rush job with so many uncertainties. Barton sensed Butler's irritation. The following week he asked Beebe and William Gregory of the American Museum to write letters to Cox & Stevens encouraging them to finish the work within six months.

Beebe wrote on the letterhead of his Bermuda Oceanographic Expedition:

> This is simply to tell you how keen I am about Otis Barton's deepsea tank, which I hope to help him with in Bermuda this summer. I am glad to provide the large *Arcturus* winch, & wire if necessary, & have arranged for a satisfactory boat in Bermuda for the work.

Gregory wrote from the Department of Ichthyology of the American Museum:

> In connection with Dr. Beebe's expedition to Bermuda, I am especially interested in Mr. Barton's plans for a deep-sea diving sphere.
>
> I trust that you may be able to push this work forward so that we can begin testing it in Bermuda this summer.

So Butler dove in and started lining up subcontractors for casting the sphere, winding the hoisting cable, and supplying the telephone and electric lines. In any field, designers and builders must experiment, fail, and correct their mistakes. But Butler had no reports or journals on deep-diving craft for references, no previous failures from which to learn the limits of the envelope in which they were operating. Even after test dives of the empty sphere, a manned descent could result in the unspeakably horrible deaths of William Beebe and Otis Barton if any one of three primary components failed. The sphere, its portholes, and its hatch had to remain intact and absolutely watertight; the cable had to hold;

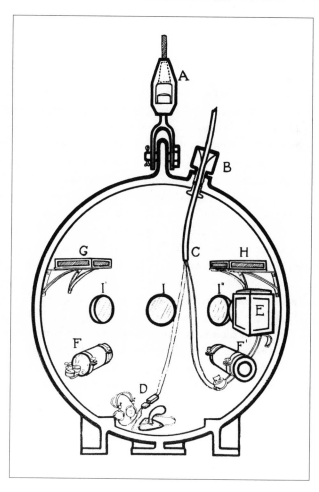

A. Clevis
B. Stuffing Box
C. Phone/Electrical Cable
D. Telephone Headset
E. Searchlight
F. Oxygen Tanks
G & H. Chemical Trays
I. Portholes.

and the air inside had to be breathable from the moment the hatch was sealed until it was opened again on the surface.

Butler took the blueprints to Watson-Stillman Hydraulics, a company with a plant in Aldene, New Jersey, that specialized in heavy cast-iron and steel housings for industrial hydraulic presses and cutting machines. William Waters, the chief engineer at Watson-Stillman, was intrigued by the challenges of casting Butler's diving tank and grateful for a new cash customer, but he knew the job would be dicey. He would pass on the actual steel casting to the Atlas Foundry in Buffalo, New York, and oversee the job and finish the sphere in New Jersey. Atlas usually made machine parts that could be cast in sections and riveted together, and the company name was familiar to factory workers around America who saw it stamped into belt housings, boilers, heavy cutting tools, and presses. Casting a hollow steel sphere in a single pour, however, was a very different proposition. The sooty crew of pattern makers, stokers, and blacksmiths would have their hands full, especially since Barton and Butler wanted their sphere cast by May and finished in the Aldene, New Jersey, machine shop by June. In 1929, Atlas, Watson-Stillman, and every other tool shop, foundry, and mill around New York City were already running at full speed to keep up with the greatest steel and iron building boom in the history of the world. But Waters said he'd give it a try. A week later, he sent Butler an estimate of $2,300 and a plan for casting the sphere.

None of the Atlas Foundry's pattern makers, who would actually shape the casting, had ever built a diving sphere, but they quickly figured out that they could use the same techniques that had been proven by centuries of casting church bells. They would form a plug of damp sand and clay roughly the size of the sphere, bake it, and file it smooth to the precise dimensions of the interior walls. Then they would improvise, forming the hatch combing, the porthole housings, and the lug to which the vital hoisting cable would be attached. Next, they would surround the hardened sand globe with steam-bent wood in arcing sections built up in layers to the two-inch thickness specified for the walls of the sphere. Over the wood, they would form sand and clay into clamshell halves and bake them. The entire assembly would consist of the solid inner core of hard, baked clay, two inches of wood, and the two halves of the baked clay clamshell on the outside. The pattern makers would carefully remove the outer clamshell mold and the wooden interlay, and then reassemble the outer mold leaving a two-inch void between it and the central plug. The blacksmiths would then pour molten steel into that void to form the diving sphere.

Steel was ubiquitous by the twentieth century, a metaphor for strength and durability and the seminal icon of the Industrial Revolution. Mills and foundries observed highly evolved manufacturing protocols, but the metal itself was subject to corruption by air, error, and contamination; steel failed every day, buildings and bridges collapsed because flaws in chambers, beams, girders, and rails had gone undetected in foundries. For Barton's sphere, molten steel at 2,500 degrees Fahrenheit would be poured rapidly into the void created by the pattern makers through openings called sprues, with the air escaping through other holes called vents. When the steel cooled, the outer mold would be removed, and the inner plug chipped out through the hatch opening. The rough casting would then be taken to another shop where grinders and polishers would finish the steel and inspect it for hollows, chinks, and other flaws that could send the job back to the pattern makers, who would start over.

Leaving Watson-Stillman and Atlas in charge of the actual casting, John Butler turned to the second critical element of the diving craft: a cable to lift and lower the five-ton sphere that wouldn't break, kink, or twist badly enough to prohibit rewinding it onto a winch. Butler knew that if he was building a vessel in which to lower America's most famous naturalist and his young partner from New York into the depths of the Atlantic Ocean, there really was no choice but Roebling's.

By 1929, John A. Roebling's Sons had achieved the kind of celebrity as a business enterprise that is usually reserved for movie stars and athletes. The company had been founded by John Roebling, a German immigrant who came to America in 1831 to farm in Pennsylvania and quickly figured out that irrigation and transportation were central to his success. Roebling, who had studied at the Polytechnic Institute in Berlin, took a job as an engineer on the Pennsylvania Canal then being dug in the foothills of the Allegheny Mountains to move water to the farmers, coal and iron ore to the mills, and finished products to market. At several points along the route of the canal, the barges had to be portaged up hills too steep to excavate, a matter of dragging them on greased skids with hemp hawsers nine inches in diameter. The thick hawsers were expensive, hard to handle, and often broke under the strain of the enormous loads. Rope of that kind was made by winding natural fibers into bundles and then further increasing their strength by braiding them in continuous sections on long covered platforms called rope walks. While keeping up with technological developments in Germany, Roebling had come across a paper describing the production of continuous lengths of iron in a process called extrusion. Molten iron with the right balance of carbon and manganese could be forced under great pressure through small, hardened steel channels to make flexible wire.

John Roebling left his job on the canal, went back to his farm in Saxonburg, ordered a shipment of wire from Germany, built himself a rope walk in his meadow, and taught himself to braid steel wire into cables. A year later, he succeeded in producing a flexible wire rope one and a quarter inches in diameter that was as strong as a nine-inch hemp hawser and would last many times longer. Roebling was in business. After initial skepticism, the Pennsylvania Canal switched all its portaging hemp to his wire rope. Then Roebling saw an even greater opportunity. The canal had to cross natural rivers on its way across Pennsylvania, and the only way to do that was to construct aqueducts over them in which the canal would flow. That was standard practice, but the aqueducts had to be made of stone, took a long time to build, and were usually the most expensive parts of a canal. With the canal poised to cross the broad Allegheny River, Roebling persuaded the state to let him build a revolutionary new kind of bridge that depended upon his wire rope instead of rock for suspension. The wire-suspension aqueduct was inexpensive and plenty strong, and Roebling never looked back. Three years later he built the first suspension bridge across the Monongahela River at Pittsburgh, then a dramatic span across the Niagara River Gorge that made headlines and rocketed him and his company to fame. Roebling suspension bridges would eventually cross New York City's Hudson and East Rivers, and hundreds of other waterways all over the world. John Roebling fell ill and died in 1869 as a result of an injury sustained while building the Brooklyn Bridge, which was finished by his son, Washington Roebling.

The mines and mills were churning out rails, beams, and girders, and buildings with steel skeletons were rising higher and higher into the sky, but the ingredient that made the skyscraper practical was the elevator pioneered by the Otis brothers. They hung their elevator cars with counterbalanced rigging and safety brakes, and they bought all their cable from the Roeblings. By the time Butler ordered the cable for Barton's diving sphere, more than ten thousand people worked at John A. Roebling's Sons & Company under its president, Ferdinand Roebling, Jr., John's grandson. The founding family still owned the company, with workers who were fiercely loyal to their beneficent employers. Roebling's set up generous pension funds, paid well, and gave to charities and good causes, including the work of William Beebe and the New York Zoological Society.

Butler wrote to L. W. Bennett at Roebling's, with whom he had worked rigging ships, and asked for advice and a cost estimate for six thousand feet of wire rope capable of lifting a five-ton steel ball and supporting it and the weight of the cable at a depth of one mile. Bennett told him that a standard three-quarter-inch, nontwisting, Blue Center steel rope with a breaking strength of twenty-five tons would work just fine and had been proven reliable as elevator cable. The nontwisting variety, Bennett explained, was wound with alternating strands in opposite directions around a solid steel core to resist tangles and kinks and still remain very flexible, which was exactly what Butler thought he needed for a deep-diving bell. Bennett could guarantee delivery three weeks after getting a firm order at thirty-one cents per foot, with a discount of from 5 to 20 percent depending on how quickly the bill was paid.

Butler stewed over the breaking strength of the cable, awash in doubt because he wasn't convinced that static breaking tests were reliable for proving the cable's ability to carry the weight of a steel pendulum swinging in the ocean a half mile below the surface, buffeted by uneven strain from the motion of the mother ship and ocean currents. Eventually, he increased the specifications to a diameter of seven-eighths of an inch to gain an extra four tons of breaking strength. Roebling's told him the thicker cable was not in its standard inventory, would have to be specially spun, and would cost twice as much as the three-quarter-inch. But did he really need six thousand feet of it? Butler talked to Barton, who admitted that a reasonable expectation for the deepest dive was a half mile instead of a mile, so they ordered 3,500 feet of the thicker cable. The $1,500 cost was the same as Barton would have paid for six thousand feet of the three-quarter-inch, and the extra eighth of an inch of diameter boosted the breaking strength to twenty-nine tons. Butler rested a little easier.

Butler's work on the design for the electrical and telephone systems was no less freighted with the specter of potentially dire consequences in the event of failure. The divers inside would breathe from tanks of extremely flammable pure oxygen; a single spark could incinerate them in a heartbeat. Barton's initial vision of his deep-ocean craft included lamps mounted on the outside of the sphere, which meant that a water-proof electrical cable would have to run from the surface to the separate exterior housings for the lights. Barton also wanted a telephone line, the cable for which would have to enter the sealed sphere in some way. The technology for producing electrical and telephone lines that could with-stand immersion was readily available in 1929, and heavily armored transoceanic cables were already in common use. Butler wasn't at all sure, though, how a thin power transmission cable would stand the pressure of great depth or the strain of a descending diving tank. General Electric was the obvious source for the information Butler needed to draw up his specifications, and in early March he wrote to W. L. Enfield at the GE office in Cleveland that specialized in lamps:

> We have designed, and are about to have built, a deep sea bell of which the plan is enclosed herewith.
>
> This bell is to be used by a nationally known scientist [Otis Barton] this summer in an enterprise which will be an official undertaking of the American Museum of Natural History.
>
> It is to be made of cast steel and capable of descending into the ocean to a depth of one half mile; that will mean a pressure of 1175 pounds per square inch.
>
> It will be lowered from the "mother ship" and will be equipped with a port hole through which observations will be made.
>
> You will appreciate the many problems we have had to cope with, among them being the type of lamp that will be practical.
>
> Last week I had the pleasure of meeting [glass expert] Dr. E. E. Free, 175 Fifth Avenue, New York City, who showed great interest and advised me to write to you explaining my problem.
>
> Considering the fact that man has never been below 350 feet, the experiment will be closely watched.
>
> Will you look over our idea of the placing of the light? We felt that inasmuch as a light placed outside as shown would have to be encased in a casting the same thickness as the bell itself, it would be much simpler to place the light inside.
>
> Again, as the bell is rather small we wonder if the heat generated by the lamp would be sufficient to cause discomfort to the scientist inside

who would be absolutely sealed for a space of say two hours including the time to lower and raise the ball. In each case the feed lines are carried up to the "mother ship."

Dr. Free felt that as the nature of the undertaking is highly scientific you would be able to advise me in many ways.

An extract from a report regarding deep sea operations says: "the zone of effective illumination (by the sun) extends only 600 feet down. Man has not descended to what we call the preliminary abyss. Below 600 feet the 'twilight zone' is similar to pale moonlight. The new television ray discovered by the Scot, Baird, might possibly be of use in bringing close to man's eye the conditions in which submarine fauna live."

Our question would boil itself down to the type and power of the lamp to use. Having no data I would hazard a guess that the range of vision would be about ten feet.

We will also be grateful if you will give us an idea of the cost of the installation and the time to manufacture the lamp from the date of order.

Dr. Free advises me to run out and see you. In view of the pressure of work at our office I hesitate, unless you think it's absolutely necessary.

Enfield passed Butler's letter on to another General Electric engineer, C. E. Egeler, who replied two weeks later:

. . . From our experience with lighting for divers and from available information on the problems of underwater illumination, I feel that it will be advantageous to place the lamp inside the deep sea bell behind a window, for three reasons: First, the lamp will be accessible in case of burn-out; second, the heat radiated inside the bell will probably improve the comfort of the observer or photographer—also, it should reduce the condensation of water vapor inside the bell; third, while we have made lamps which would stand several hundred pounds water pressure, less possibilities for trouble would obtain if a lamp were used behind a window, since the features of water-tight electrical connections and the development of a new lamp to withstand extremely high pressures would be avoided.

. . . We would recommend that provision be made for a 1000-watt 115-volt G-48 bulb spotlight type lamp with reflector to be placed behind a window approximately 6 inches in diameter, although 8 inches would have some advantage.

Butler welcomed the advice he received from the General Electric engineer because it simplified his design. He no longer had to consider a separate housing and transmission line for the searchlight and could place the electrical and telephone circuits in a single waterproof tube that would enter the sphere through just one pressure-proof hole. Calculating the amount of electricity required to power the lamp was straightforward. Barton told him he had the 5-kilowatt generator from William Beebe's research ship *Arcturus,* which could produce 110 volts at 45 amperes. He knew that the half-mile-long size 8A copper-coated steel wires through which the electricity would run would absorb a lot of that power, but there would be enough left to run the lamp. There were still a lot of details to work out, but Butler was confident that he had a good plan for the light.

For the telephone system, Butler wrote to C. R. Moore at Bell Laboratories and was relieved to learn that communications between the sphere and the surface could be handled with off-the-shelf components. Moore prescribed a pair of voice headsets similar to those used by telephone operators, linked by a pair of size 14A copper-coated steel wires and powered by a 22.5-volt radio battery on the deck of the mother ship. He and the other engineers at the company founded by Alexander Graham Bell were so impressed by Barton's ambitious plans that they gave him the whole system at no cost.

Once Butler settled on the specifications for the electrical and phone lines, he had to find a way to protect the cables from breaking under enormous pressure. Without lights the divers wouldn't be able to see the inhabitants of the abyss, and without communication the deck crew wouldn't know if the divers were even alive. He asked General Cable and Westinghouse for bids, but both companies said they weren't equipped to produce a thin, armored, waterproof hose containing four wires a half mile long. An engineer at Westinghouse, though, told Butler that there were three smaller manufacturers who might take on the job, Simplex Cable of Boston, Safety Insulating Cable of Bayonne, and Okonite Cable of Passaic. Butler sent out his barrage of letters and quickly heard from Simplex and Okonite. The Simplex proposal, despite the company's name, was a complicated combination of conductor wires for the electricity and phone lines, heavy rubber sheathing, jute and tar packing, and steel armor. Okonite, on the other hand, had recently patented a process for heat-sealing wires of almost any size in a strong, pliable rubber tube.

Butler liked its simplicity. The Okonite power and phone line would contain the four transmission wires in a ¾₄-inch-thick tube of their invention, Okonite, surrounded by another patented sheath of ½₂-inch Okocord with a total outside diameter of 1.115 inches. It would be extremely flexible and could be spliced if a longer length were ever needed.

Butler had his electricity and telephone cable. The problem then was getting Okonite's revolutionary new cord through the steel wall of the sphere to the phone and lamp inside. He turned to a device upon which seamen had depended for decades. A hole in a boat is tantamount to disaster, so when marine architects and shipbuilders were confronted with the challenge of running a turning propeller shaft through a hull they had to reconcile an ancient fear with the immense advantages of a power plant. In the beginning of the engine era, a propeller shaft was simply passed through a hole in the hull cut to tight tolerances, but the cast-iron shafts of those days ground the hole larger with every turn. Mariners and shipbuilders tried packing the shafts inside and outside the boat with rags or rope, but that didn't work too well either. Eventually they came up with replaceable boxes made of a pair of hollowed-out blocks set into the hull, packed with jute and tar, and clamped together. The stuffing boxes, as they were called, leaked, but not enough to sink a boat, and the seepage actually helped to lubricate the shaft. Steel-hulled vessels presented another set of problems when they came along, but the concept remained intact and the name the same. A stuffing box for a propeller shaft passing through a steel hull consists of an outer fitting and an inner fitting threaded together with a soft packing of hemp or oakum between the two fittings. When the threaded pieces are tightened together, the packing compresses to seal off the inside of the hull from the sea.

No one had ever used a stuffing box to keep out water under more than a thousand pounds of pressure per square inch, but Butler had no other choice. There had to be a hole into the sphere or there could be no lights or telephone. Butler specified a three-inch stainless steel nut for the outer fitting and a similarly sized brass nut for the inner, through each of which would be drilled a hole three millimeters larger than the Okonite tube. The threads of softer brass would seat firmly against the harder steel, and between these fittings, a packing of oiled flax would create a seal around the cable passing through them. The inner brass nut could be tightened with a wrench from inside the sphere if it began to leak, squeezing the oiled flax more tightly into the gaps. The stuffing box should work, and if it didn't, he would find out in the first unmanned test dive, which wouldn't kill anybody.

By mid-February 1929, Butler was confident that Barton's tank and its cable would be strong enough to withstand the pressure of a half-mile dive, and that it would have lights, power, and communication with the surface. The glass through which the divers would peer into the abyss, however, remained a puzzle. Butler wrote to Willard Morgan at Triplex Safety Glass of Clifton, New Jersey, which supplied glass to the navy for the eye ports of submarine conning towers. Morgan told him that the submarine ports were made of ordinary plate glass three inches in diameter and three-quarters of an inch thick, and had withstood the pressure down to a depth of 250 feet. He said that plate glass of that size and thickness had been tested under pressure of eight hundred pounds per square inch, which meant it would hold down to a depth of about 1,700 feet—but no more. He had not tested laminated glass, but doubted it would be much stronger. He also pointed out that increasing the size of the glass to the eight-inch diameters required for the diving sphere increased its weakness exponentially, so he couldn't recommend it. And because of the impurities in plate or laminated glass, if they increased its thickness enough to withstand the pressure, it would be too opaque to see through.

Another piece of information cast a shadow of deepening pessimism over the diving sphere. During his research on glass, Butler had come up with a report on trawling at great depths for which airtight, hollow glass balls were attached to the fishing nets to hold them open. On the surface, the balls were perfectly impervious to water and never leaked. When they were brought up from a deep dredging operation, though, some were full of water, and even after the ball had been on the surface for days, the water had not evaporated. Butler wondered whether something had happened in the abyss to transform the glass and actually allow water to pass through.

Butler went to E. E. Free, who told him not to worry about the water in the glass balls, which had probably gotten in through microscopic holes. Free also told him that he had to find someone to build windows of fused quartz, a revolutionary kind of glass made from quartz sand— pure silicon dioxide—subjected to temperatures in a vacuum furnace high enough to force out any air bubbles that might cause holes. Quartz glass was first produced in France in 1839 by a chemist named Gaudin, who succeeded in making small beads in a very hot flame. The old adage that what was genius the first time is soon managed by a tinsmith did not apply to making quartz glass. A practical method for manufacturing it did not become possible until eighty-five years later, but just in the nick of time for Butler, Barton, and Beebe.

Quartz glass would be strong enough to withstand the water pressure on an eight-inch disk three inches thick at a depth of three thousand feet, Free said. The bad news was that pieces of quartz glass of the size needed for even the most minimal observations of the deep ocean from a diving sphere were bigger than any that had ever been produced. The only company with the technology to do it was General Electric, where in 1924, after fifteen years of research and the invention of special furnaces and vacuum chambers, engineers had successfully manufactured useful quantities of quartz glass. With it, General Electric dominated the market for lenses, prisms, mercury vapor lamps, insulators, and a remarkable new safety device, the fire sprinkler head. Quartz glass could be perfectly calibrated to disintegrate at a specific temperature, and GE sold millions of their sprinkler heads during the skyscraper building boom in the twenties. Free told Butler that quartz glass would be perfect for underwater observations because it transmitted the full spectrum of light without perceptible distortion. And, most important, it could withstand a pressure of seven thousand pounds per square inch.

A week after he met with Free and considered his recommendations, Butler told W. H. Jones at General Electric in Schenectady to make each of the three pieces of quartz glass three inches thick and precisely 7.965 inches in diameter to fit the window housings in the sphere. By the middle of April, Jones phoned to say that General Electric had succeeded in making the windows, which would be polished and delivered to Watson-Stillman by the first week of June. Butler had done it. The sphere, equipped and ready to dive, would be shipped to Bermuda by July 1.

Five

BERMUDA

*The job I had set myself on my new island home was
the study of the fish of the deep sea and the shore.*

William Beebe, *Nonsuch: Land of Water*

While Otis Barton and John Butler were spending the spring of 1929
absorbed in the increasingly fine details of building something
nobody had ever seen before, William Beebe was in paradise, where some
good luck had brought him an enormous windfall. In the autumn of 1928,
just before Barton had come to the Bronx Zoo with his blueprints and sold
him on the design for a diving tank, Beebe had gone to Bermuda for three
weeks with Elswyth, his assistants Gloria Hollister and Ezra Winter, and
his capuchin monkey, Chiriqui. The trip was a reconnaissance to investigate the possibilities of doing some fieldwork in the midocean archipelago,
and Beebe spent most of his time wading in tide pools, dining, dancing,
lecturing, playing deck tennis, and wielding his charisma to charm the
cream of Bermuda society. He was immediately placed on the A-list, which
that season also included His Royal Highness Prince George of England, a
visitor to Bermuda aboard the cruiser HMS *Durban*. On October 24
Beebe went with Hollister to a garden party where he shook hands with
the prince, who told Beebe that he had read his new book, *Beneath Tropic
Seas,* the account of his last expedition to Haiti. Was there any chance,
Prince George asked, that Beebe might take him diving?

Beebe could barely contain himself. His own life had taken a sharp
and fateful turn three years earlier when he had stepped down a ladder
into the water of Darwin Bay in the Galápagos wearing a copper diving
helmet. "Tried helmet diving for the first time, and found it a most exciting experience," he had written in his diary. "Trite but true to say it
opens a new world." Beebe named his new undersea realm "The King-

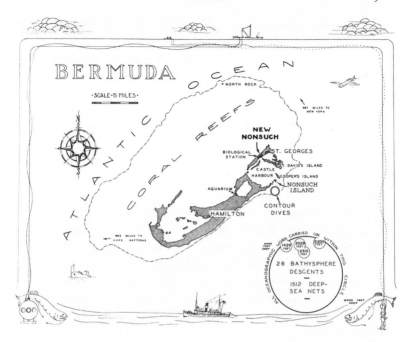

dom of the Helmet," embraced walking around on the bottom of the ocean with the enthusiasm of a kid turned loose in a toy store, and became an unabashed proselytizer for helmet diving. "The only requirements are a bathing suit and a pair of rubber-soled sneakers, a copper helmet with glass set in front, an ordinary rubber hose, and a small hand pump," he wrote in one article. "Down you go into two, four, six, eight fathoms, swallowing as you descend to offset the increase of pressure."

Beebe churned out advice to his readers on scientific observation, underwater photography and moviemaking, spearfishing, and even the creation of shoreside undersea gardens through which you could take your dinner guests for a stroll much as you would through a patch of prize-winning roses. He actually thought that every beachfront home someday would have its own underwater garden.

> If you wish to make a garden, choose some beautiful slope or reef grotto and with a hatchet chop and pry off coral boulders with waving purple sea-plumes and golden sea-fans and great multi-colored anemones. Wedge these into crevices, and in a few days you will have a sunken garden

in a new and miraculous sense . . . Our grandmothers lined their garden paths with conch shells, but under-sea it is more difficult to do this, for the giant snails will insist on walking away as soon as you have planted them . . . Finally, as a border to your marine plantation, collect a score of small, rounded brain corals all thickly covered with tube worms . . . All are in motion, though there is no current, and we feel that there would be nothing remarkable in their suddenly saying, like Alice's Tiger-lily, "We can talk, when there's anybody worth talking to."

Helmet diving captured the public's fascination. Underwater, divers salvaged ships, built things, played musical instruments, painted pictures, set endurance records, and competed in races. In 1927, Gustav Kobbe set the world record for submarine speed walking when he covered five miles eight feet down in the East River from 150th Street to midtown Manhattan in two hours and twenty-seven minutes, beating the time of William Smith, the champion English underwater walker.

Just before he returned to New York in November 1928, Beebe topped off his wildly successful trip to Bermuda by going helmet diving with Prince George.

> *Began to pack today and decided to leave many things behind as it seems worth while to return next year for 6 months to work up the fish, etc. for a Bermuda book and for life histories.*
>
> *Commander Shelly and HRH Prince George came at 11 a.m and although it was rather rough we went around Kings Point and out to one of our favorite reefs. I went down after we had a time anchoring the big cutter, then PG descended. He was out 20 feet and down 12 when we began to drift and the stern with the ladder drifted over part of the reef. I was terrified lest he should not realize the conditions and should try to reach the ladder at this moment, but luckily as I was about to dive and climb along the hose and lift it off him, Elswyth called out that he was on the other side of the boat. I got him back and by this time we had shifted away from the reef. He came up the inside of the ladder and I got him and all was well, but for five minutes we were the most worried group of people in the world.*

The day after he rescued Prince George, who thought the whole diving adventure was "just bully," Beebe went on a farewell outing with

Governor Louis Bols and his entourage to Nonsuch Island on the southern rim of Castle Harbor. The island had been uninhabited since a yellow fever quarantine hospital there closed a few years earlier, but it still had serviceable buildings and a boat landing. While the island caretaker and his wife set up a champagne picnic in a central clearing, Beebe, Governor Bols, and the others traipsed around its twenty-five acres through Bermuda cedars, St. Augustine grass, and lantana scrub. They looped around from the hundred-foot bluffs of the western point overlooking Castle Rocks, past a freshwater pond, a saltwater marsh, the tidepools on the eastern shore, and the sandy arc of South Beach on a perfect pocket bay. By the time they reached South Point, with the purple blue of the deep Atlantic in front of them, Beebe had fallen in love with the place. His enchantment wasn't lost on Governor Bols, and as the party was walking back to their boat, he startled Beebe by offering him Nonsuch as a permanent base for the oceanographic research of the New York Zoological Society. Bols liked Beebe, but his affection alone was hardly enough to motivate such a gift. For Bermuda, the presence of a popular naturalist from New York would be a boon to tourism and to the reputation of the remote little colony, because every account of his discoveries and exploits would carry its name. In return, Beebe would have a comfortable, safe research station with a full-scale scientific laboratory in one of the best places on earth for studying the ocean. Beebe accepted Bols's offer on the spot.

Back in New York, Beebe sold the idea of the Nonsuch field station to friends and patrons, who wrote checks to pay for the first of what he hoped would be many seasons of deep-sea exploration of a submarine cylinder two miles deep and eight miles in diameter, nine and a quarter miles south-southeast of his island. Harrison Williams and Mortimer Schiff together contributed $15,000. Marshall Field gave Beebe $2,000, George Baker $1,000, Coleman duPont $2,500, and Bill Boeing $1,000, and others brought the total to $28,500 in cash. Beebe himself added $1,000, and the Bermuda Board of Trade kicked in $500. Others donated equipment, books, and supplies. Roebling's gave Beebe's expedition 15,000 feet of trawling cable, George Eastman 5,000 feet of movie film, and Carl Zeiss sent a telescope that became one of Beebe's personal treasures, with which he was to end many evenings on the western bluff of Nonsuch peering into the cosmos.

The First Bermuda Oceanographic Expedition of the Department of

Tropical Research of the New York Zoological Society arrived on Nonsuch Island on March 15, 1929, with ninety crates of nets, cables, jars, aquariums, pumps, generators, microscopes, thermometers, formaldehyde, and everything needed to set up housekeeping on Nonsuch. Beebe's staff included John and Helen Tee-Van, field assistant William Mirriam, artist Ellen Rogo, photographer Sven von Hallberg, and lab assistant Phyllis Boyden. The expedition's chief technical associate in charge of preparing specimens was an ichthyologist named Gloria Hollister, who, not incidentally, was also at Beebe's side for dancing, tennis, and parties whenever Elswyth wasn't on the island. Hollister was five eleven with gossamer blond hair cut short around a face that broke easily into a toothy smile. Everybody called her Glo. She was the daughter of a New York doctor who had introduced her to nature at the family farm in the Ramapo Mountains of Rockland County, New York. Her father was devoted to her, and joined in her teenage hobbies of raising prize poultry, dogs, cats, and horses, for which she won blue ribbon after blue ribbon in competitions and exhibitions in New York, Ottawa, and London. In 1927, flocks of her Black Orpington and Naked Neck chickens were shipped across the North Atlantic to introduce those breeds to Russia.

Hollister had gone to Miss Rayson's in Manhattan and the Hillside School in Norwalk, and in the fall of 1920 she had enrolled as a zoology major at Connecticut College for Women in New London. She had been a precocious science student, but also played varsity basketball, field hockey, and soccer, and was a high jumper and discus thrower on the track team. She was on the All-American girls' field hockey team that met the world champion team of Great Britain in the fall of 1920 and suffered a humiliating defeat. "Those nimble, short-legged British girls just ran rings around us long-legged Americans," she said when she told the story.

After graduating at the top of her class in 1924, Hollister enrolled at Columbia as a graduate student in zoology under Florence Lowther and William Gregory. After three years of work in a medical laboratory, in the spring of 1928 she heard that William Beebe was organizing an oceanographic expedition and needed a professional naturalist to handle dissection and preparation of fish. Hollister asked Gregory to introduce her, and Beebe hired her the day they met. Six months later, on the first trip to Bermuda, when Governor Bols gave Beebe Nonsuch Island, Hollister fell in love with her boss. In hieroglyphic code in his diary, Beebe recorded the moment when he and Gloria were exploring a cave and she kissed him for the first time. Beebe and Elswyth had a modern marriage, an arrangement common in avant-garde circles in New York, in which neither of them was obligated to remain sexually exclusive as long as their life together wasn't damaged. The Beebes easily kept up appearances, and since Elswyth hated hot weather, she spent very little time on Nonsuch Island. There, Gloria Hollister was the alpha female.

Beebe's expedition was already in full swing on June 1, 1929, when Otis Barton sailed from New York with John Butler's promise that the diving sphere would be shipped to Bermuda in early July. Barton had made the two-night voyage aboard the Furness liner *Fort Victoria* before, but never with the anticipation that accompanied him as he watched the skyline fade against the afternoon sun and the ship settled onto its southeasterly course. He sailed in first class, enjoying the cocktail hour at dusk, dinner in good company, and a walk on deck, but when he looked over the stern rail at the blue-black Atlantic mending itself in the foam of the ship's wake, the audacity of his intention to descend into the abyss overwhelmed him. So much of his life had been laced with half-baked adventures that left him with no sense of real accomplishment, and now he was sailing off to try something no one had ever done before. Later, during a sleepless night in his stateroom, he felt more like a schoolboy waiting to take an exam than a man poised to fulfill a dream. After months during which most of the details of bringing his dream of a diving sphere into reality had been handled by Butler, Watson-Stillman, General Electric, Okonite, and Bell Labs, Barton had real work to do. It would be up to him to rig a tugboat with the winch from Beebe's *Arcturus* and install the drum of cable from Roebling's. Butler could help with the engineering and load calculations by mail or telegram, but he had already warned Barton that he wasn't sure the *Arcturus* winch could lift the five-ton sphere and its three tons of cable without some ingenious rigging.

On the second morning of Barton's voyage, the *Fort Victoria* was six hundred miles out of New York, grumbling and bucking in a young northeast gale. He spent some time on deck, feeling queasy and watching the low brown eyelash of Bermuda rise on the horizon. The several hundred islands, most of them small and unnamed, are no higher than fifty meters, composed of an Aeolian limestone crust over the top of a volcano that erupted on the ocean floor a hundred million years ago. Seen from the air, the island group is shaped like a fishhook, with its shank beginning in the northeast and running fifteen miles southwest to its crook, which forms Great Sound, one of the world's legendary anchorages. The big harbor was among the most prized possessions of the empire when the British controlled over one-third of the people on earth, and Bermuda was to the Atlantic what Gibraltar was to the Mediterranean.

The *Fort Victoria* navigated through the northern shoals into Great Sound to call at Hamilton, the center of government, commerce, and colonial society. Tee-Van met Barton, helped him get his luggage on the train to St. George, and took him to Nonsuch, a half hour through sheltered water in a skiff with an outboard motor. On Barton's first day on the island, Beebe treated him as a guest and took him on the standard nature walk through the cedars to the bluff, around the pond and salt marsh, and out to Southeast Point. The next morning they went diving. On the boat ride out to an offshore reef, Beebe pointed out dents in his metal helmets made by lava blocks in the Galápagos and scratches from the caves off Panama, which, to Barton, were even more awe-inspiring than the specimen jars, maps, and photographs in Beebe's office the first day they met at the zoo. Barton followed Beebe down the ladder to the bottom, and he could hardly believe where he was. Thirty feet above him, the red keel of the launch rolled on a gentle swell, right in front of his faceplate, a school of sergeant majors swam by, and beside him stood William Beebe, pointing his shark spear at a pair of cruising barracuda in the distance. On the ride back to the island, Beebe kept up a steady patter of names and observations, thrilled as always by what he had seen in the world below, and Barton just listened and nodded his head, again stricken to silence in the presence of one of his heroes.

By Barton's third day on Nonsuch, though, it was obvious that it wasn't easy for a man like Beebe to have a man like Barton around. Barton wasn't interested in keeping up with the breakneck, dawn-to-dusk pace Beebe set for himself and everyone who worked for him, and already Beebe didn't think Barton worshiped at the temple of nature with the proper measure of devotion. Barton was good with mechanical things and fixed the generator when Tee-Van couldn't, but he wasn't a patient observer and he seemed clumsy around the trays and tanks of specimens. When Hollister was showing him around her lab, he picked up a dragon fish she was preparing and his rough handling broke off a part of its tail. Barton laughed. Later that night, Hollister told Beebe she'd just as soon not have Barton hanging around while she was working.

At about the same time Beebe began to think that Barton wasn't such good company and was not much of an asset to the expedition except for his contribution of the diving tank, Beebe's commanding presence started getting to Barton. He wasn't used to men running things, and he especially didn't like the way Beebe ignored him when he talked about anything but the tank. The glow of hero worship Barton felt for Beebe faded, the two men quickly learned to move more carefully around each other, and all hope for a spontaneous friendship was soon gone.

By the end of his first week in Bermuda, Barton had given up living on Nonsuch Island and moved to the St. George Hotel, telling Beebe that he just wanted to be closer to the wharf and his work on the tug. In truth, he much preferred the old town with its narrow streets and easy evenings on the veranda to trying to keep up with Beebe. St. George, named after the dragon slayer and patron saint of England, was founded in 1612, making it the oldest continuously occupied English town in the New World. It was Bermuda's capital for more than two hundred years until the government moved to Hamilton in 1815.

Barton learned right away that Bermuda was no place for a rush job. Nothing happened quickly in the tropical outpost, where no one but doctors and senior members of the government were allowed to have automobiles and trucks. Horses pulled carts and drays, and most freight and passengers moved by boat or the Bermuda railway, which made the fifteen-mile run between Hamilton and St. George along the north coast in about two hours, stopping at stations and flags along the way. Fewer than 25,000 people lived in the archipelago, most of them descendants of the survivors of wrecked slave ships or British colonials, and the rhythms of life were languorous in a charming way, often depending simply on whether it was raining or not. Though Bermuda is well north in the Atlantic, the Gulf Stream blocks the harsher currents and Arctic air of the North American continent, so the weather ranges from mild to tropical. It rains a lot, usually in the afternoon and often in torrents that shut down commerce and traffic, but the downpours are vital, since the limestone islands are porous and retain little water except in a thin fresh-water layer floating over the heavier salt water that permeates the rock. The roofs of all the buildings are plastered and channeled to capture the rainwater and direct it to cisterns, and as Bermudians wait out a squall, they take pleasure in knowing their water supply is being topped off.

Barton set up shop at the Darrell and Meyer yard, which had been repairing ships for more than three hundred years. The *Gladisfen,* the vessel Beebe used for trawling, had broken down, so he chartered another veteran tug, the *Freedom,* and Barton went to work figuring out how to lift the sphere from the deck and then manage it and the cable in the water. Marine architects have been calculating loads, winch capacities, and block-and-tackle systems for centuries to design anchors, chains, and rigging, but there were some new puzzles involved in the behavior of a hollow steel ball at the end of a cable.

After Beebe told Barton that he wanted to use the winch from the *Arcturus,* Butler had written to Ernest Pulsford of the Lidgerwood Manufacturing Company in Elizabeth, New Jersey.

> The question at issue we understand to be whether or not our #1917 winch geared to the reel drum as furnished by us will have sufficient hoisting capacity to hoist the bell weighing 10,000 pounds a sufficient distance above the deck to lower it into the water . . . It is a little doubtful as to whether there would be sufficient pull with all the cable on the reel to lift the bell of[f] the deck of the steamer and lower it into the water, but this difficulty could be obviated by using a 3-part purchase operated by the main drum of the winch. When this is done, sufficient slack could be pulled off the large reel, and when the bell was lowered into the water the 3-part purchase could be disconnected and the bell then lowered by the 1″ cable on the reel.

The *Arcturus* winch wasn't powerful enough to pick up the five-ton sphere from the deck when all the cable was on the drum because its power would be at its lowest point, in the same way that moving a chain over a large bicycle sprocket at the pedals requires more leg power than moving a chain over a small sprocket. Maybe, Pulsford suggested, they could use block and tackle along with the winch to get the sphere off the deck and into the water. Once the sphere and cable were in the water, they would be buoyant and lighter, so the winch alone could handle the load.

For a few days at the end of June, it looked as though Barton had the weight problem licked. Just to be sure, he drew a sketch of his rigging system and mailed it to John Butler back at Cox & Stevens on June 25.

I have just talked with the chief of the *Freedom*. He believes the following set up would be possible. [Here the sketch] The cable goes from the tank to the block on the boom, down thru a block on the deck, forward to the big [*Freedom*] winch around the drum of which it passes five or six times; back to the reel and winch of Mr. Beebe. Thus both winches act at once.

What do you think of this?

Photos of the *Freedom* are printing now at Abercrombie & Fitch.

The same day, however, Captain Sylvester decided that the mast of the tug was just not strong enough to carry the load. No matter how many turns are taken through a block, the full weight of a lifted object eventually must be borne by the point at which the block is attached to

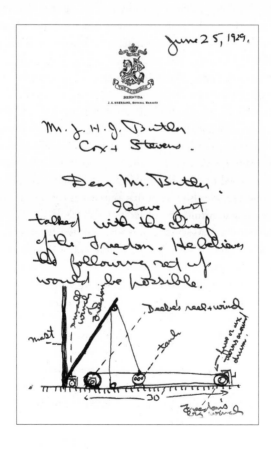

the mast. Sylvester called a halt to the rigging experiments on the deck of his tugboat and Barton wrote another letter to Butler.

> Since writing to you today I have again talked to Captain Sylvester. He believes that the double winch idea of my previous letter is not feasible. He is doubtful about [the strength of] the boom and mast of his boat.
> I believe therefore that we must give up the idea of the *Freedom,* unless a winch and reel suitable for the job can be chartered in N.Y. At all events, it appears to be useless to try to use Mr. Beebe's reel or winch.

By the last week of June, Watson-Stillman and the Atlas Foundry had poured a casting of open-hearth steel to near-perfect specifications, General Electric had produced the largest pieces of fused quartz glass ever made, and Roebling's cable was ready, but the miscalculation on the winch capacity killed the plan for diving in the summer of 1929. The five-ton casting of the sphere never left the Watson-Stillman shop in New Jersey.

On June 28, Barton cabled John Butler: "MUST REDUCE WEIGHT. RETURN TUESDAY. BARTON."

Barton broke the news to Beebe that his plan for rigging the *Freedom* to use the *Arcturus* winch didn't work, so diving was off for that season.

Barton went back to New York in a panic. He had come within a whisker of making a descent into the deep ocean with William Beebe and he was worried that Beebe would give up on him and go exploring with someone else. He also sensed that Butler was fed up with the project after having successfully completed the rush job on the diving sphere only to find out that Barton had not worked out the details of lifting the thing. To Barton's relief, Butler surprised him on July 18, 1929, with a letter sketching out options for a sphere weighing just two and a half tons. Butler closed by reminding Barton that this time he had to figure out how to pick it up.

> We have completed calculations on the new diving Bell, and arrive at the following conclusions:
> Instead of 5'–0" internal diameter, we have reduced it to 4' 6" making the Bell 1 1/4" thick instead of 2" thick. This gives us a weight of 2 1/4 tons on deck and at the surface, (this weight is final after making ample allowances for extras).

The buoyancy of the Bell one-half mile beneath the surface water will be about 1.4 ton. The weight of 3000 feet of rope (when bell is submerged) would be about 1.35 tons.

Can we arrange for this weight? In justice to you and to ourselves we cannot further decrease the thickness of this casting. Although we have not as yet heard from Lidgerwood (these calculations take time, especially when there is other rush work on hand) we feel that, from previous information given us Mr. Beebe's reel, as it stands, driven by his winch, would be capable of lowering and raising this weight at the rate of 100 feet per minute. That would mean that the bell could be lowered a total distance of half a mile in thirty to forty minutes, and could be raised the same distance in the same time.

So much for that. You may be asked, Mr. Barton, why ribs could not be fitted to reducing the thickness of the Bell. They would do no good, since the pressure (1175 lbs. per square inch) is constant around the bell at the designed depth, and therefore the thickness could not be reduced.

Weight could be reduced again by decreasing the internal diameter of the bell, but we do not advise this as we felt that 5' 0" was little enough, in view of the necessary fitments and accessories. We have now cut it to 4' 6" inside.

Will you kindly, at your end, make sure that this weight can be handled quite easily [Butler's underlining]. If the *Freedom*'s present boom can handle ten ton, as you informed me in your letter of June 10th from Bermuda, and Mr. Beebe's reel and winch can handle this new Bell, we cannot now see what would be in question.

Now, Mr. Barton, we regret our original Bell not being completed, but in view of the altered circumstances, believe you were justified in the change, and that despite the additional cost now, redesigning the Bell to suit the Mother Ship (instead of vice-versa) will assure economy for future operations.

Good luck, and hoping you enjoy a most pleasant vacation, we are

Very truly yours,
Cox & Stevens

Barton left for a vacation in Paris a few days after receiving the assurance that his dream was still alive. Butler went back to work on what he called Number Two. Number One was cut up and melted into ingots at Watson-Stillman, who applied a credit of $850 to Barton's account for the recovered steel against the total of $2,300 he had paid for labor and materials. Roebling's had spun the seven-eighths-inch cable on special order, and offered nothing for its return, so Barton was stuck with it even though it was much heavier than he needed for the lighter tank. The four precious fused quartz windows had never been installed in the first sphere, and they could be used in the second. The new sphere would have a smaller diameter and thinner walls, and weigh half of the original casting, but Butler calculated that the four-and-a-half-foot-diameter, two-and-a-quarter-ton sphere would safely transport Barton and Beebe to a depth of a half mile. If Atlas cast the steel according to specifications with few voids or flaws, Number Two would still be over-built by a factor of five times the strength needed to withstand the water pressure, and the cable, spun for a craft twice as heavy, would be plenty strong enough. Butler told Barton, though, that he was worried that the four-foot-six-inch interior diameter was just too small to accommodate two large men, the oxygen bottles and chemical trays, lights, and sockets. Barton told Butler that he and Beebe didn't expect to be comfortable during their descents.

Barton kept in touch with Butler by mail through the summer and into the autumn of 1929. He stayed close to home in New York, canceled a vacation to the Bahamas, and kept an eye on his stock portfolio. By October 23, the market was showing signs of finally balking after a skyrocketing climb built on margin trades that were, in essence, loans totaling over $8.5 billion, more than the value of all the United States currency then in circulation. That day, waves of panicked selling drove stock prices down and wiped out nearly $10 billion in paper value. But banks pumped money into the ruptured market and, for the moment at least, stopped the bleeding.

For Barton, the news from Wall Street could not have been worse. His failure to come up with a way to lift the diving sphere loomed larger by the hour because he might not get a second chance if he and Beebe's patrons ran out of money. The actual casting of Number Two at Watson-Stillman wasn't scheduled until early spring of 1930, so Barton had to keep Butler working and Beebe interested. On October 28, Barton wrote to Butler asking for a progress report so he could assure Beebe, who had just returned to New York, that all was well with their plans to dive the following summer. The very next day, Tuesday, October 29, another wave of selling hit Wall Street; this time no one was able to stop it before the market lost another $14 billion in value. The total loss for the week was over $30 billion, three times the entire annual budget of the federal government and nearly twice the amount the United States had spent during all of World War I.

A week before the stock market crash, though, Beebe had returned to New York with specimens, photographs, and paintings of animals no one had ever seen before. Even without deep diving in Barton's tank, the First Bermuda Expedition had been a triumph. Beebe made 528 hauls from his cylinder, collected over two hundred species of shallow-water fish, and prepared six hundred specimens for future study. Beebe gushed about his discoveries in his report to Madison Grant, his boss at the Zoological Society.

> We have blind fish, others with enormous eyes and the luminous structures are beyond words beautiful. The squids alone are marvelous, some with eyes on the ends of their arms, others covered with hundreds of scarlet and violet lights. At least a dozen fish I have not been able to assign to any known family.
>
> I am keeping notes on viability, color, shape, activities, and food, as well as eggs and colors of the lights; we are photographing and clearing them and making colored plates, so there is nothing we are not getting.
>
> The greatest interest in this work however is the continual work in one spot. We trawl from three to twelve miles off shore, southeast of Nonsuch. After the season's work is over we will be able to reproduce as nearly as is possible, the relative kinds and numbers and relationships of the deep sea inhabitants in the open ocean off Nonsuch. There is no hint of localization. The fish for the most part are of the same species found off New York and many in the Pacific.

Accurate records of what he dredged up from the deep ocean were critical to Beebe's credibility, so he and his staff had taken and developed six hundred photographs. Beebe paid particular attention to drawings and paintings, which he trusted more than photographs for scientific accuracy, and it had taken him a while to get what he wanted. He had fired his first expedition artist, Ellen Rogo, whose "style of artistic work was not adapted to my requirements," and replaced her with Llewellyn Miller, whom he judged adequate. Helen Tee-Van, John's wife, contributed some sketches and paintings, but though she was devoted and energetic, her work was mediocre at best. (Helen was the former Helen Damrosch, a daughter of New York social register parents and one of the many young men and women who found their way to Beebe as field assistants through their parents' connections.)

In July, Beebe finally found his artist when he hired a placid, talented young woman named Else Bostelmann, who, within days of her arrival in Bermuda, dazzled him and everyone else with her renditions of the strange creatures of the abyss. She had a gift for visualizing the dead or living specimens that were her subjects as though they were swimming in the deep ocean. Their anatomies signaled motion and vitality to Bostelmann, and her interpretations of them leaped from her easel. She painted the deep-sea anglerfish with its giant mouth and a light on the end of a pole rising from its head to attract prey; the silver hatchetfish, whose iridescent chrome body is lined with glowing mauve and violet lights; the gold-tentacled sea dragon; the great gulper eel; and the black swallower, able to distend its mouth and body to swallow a meal twice as large as itself.

Beebe worked hard at assigning these and hundreds of other creatures to their proper taxonomic places, but the chore would take years. He was equally excited about the results of the countless hours he and the others spent underwater in their helmets observing the animals in relationship to their neighbors and their neighborhoods. "We sought to enter the private lives of these fish," Beebe wrote to Grant, "to learn of their times and seasons, the reasons for their change of colors, their breeding habits, food and enemies."

The newspapers applauded Beebe's success as he embarked on his winter lecture circuit. "William Beebe, the man from Mars, today opened up a portfolio of drawings of things he saw on his trip," wrote George Britt in the *New York Telegram*. " 'It might as well have been Mars,' said Dr. Beebe, who is still unable to see daylight in his task of putting in order the vast collection of material with which he arrived two weeks ago. It is a task for years of study." (The title *Dr.* Beebe was correct by the fall of 1929, when that story appeared. He had been awarded honorary doctorates in the spring of 1928 by both Tufts and Colgate Universities.) Newspapers in Boston, San Francisco, and dozens of other cities played up Beebe and his weird fish in entertaining stories that were much easier to read than the accounts of crashing stocks, lost fortunes, and hard times. Reports of Beebe's wild success had the desired effect on the pocketbooks of his patrons, and despite the rising fears of economic catastrophe, he managed to raise the money for another season in Bermuda.

And Otis Barton was calling him every few weeks to tell him that he would be back in Bermuda with a new, lighter diving tank the next summer.

Six

BATHYSPHERE

Oh! hush thee, my baby, The night is behind us,
And black are the waters that sparkled so green.
. . . .
The storm shall not wake thee, nor shark overtake thee,
Asleep in the arms of the slow-swinging seas.

Rudyard Kipling, "Seal Lullaby"

Beebe loved mornings on Nonsuch Island. On June 6, 1930, he came suddenly awake as dawn flooded into his room. He swung his legs over the side of his bed and sat for a moment with his bare feet on the cool pine floor, listening carefully, hearing no sounds from the rest of the building and nothing of the wind outside. The gale that had threatened two days earlier had moved on. He heard the skrackle of longtails and drew in the salt-seaweed aromas of the ocean mingling with those of the blooming oleanders outside his window. Beebe dressed quickly in white linen shorts and a weathered khaki shirt, moved quietly past John Tee-Van's room, and then crossed the main parlor, remembering the brightness of the out-of-season fire and the good company from the night before. After a typically delicious dinner of conch chowder and baked gray snapper, he and the others had sipped rye and soda while he led them through a few tunes on his ukulele and ended the evening by reading a Kipling lullaby.

Kipling was right up there with Roosevelt in Beebe's pantheon of heroes, and his morning recollections included the fine memory of dinner with the author, Elswyth, and Governor and Lady Bols in April. Kipling liked Beebe, too, and had come to the opera house in Hamilton to hear him talk about sea creatures bizarre enough to challenge the mind of even the most exuberant fantasist. From Beebe's descriptions, it

seemed to Kipling as though the deep ocean might be as creative as a poet in its interpretations of animate possibility. In closing, Beebe told his audience that his goals included actual visits to the mysterious and dangerous realm known only to those strange creatures, a place beyond sunlight to which no human had ever gone and lived to tell about it. His associate Otis Barton, he said, had invented a craft to transport two divers safely into the abyss, and he was confident that they would descend to a depth of a half mile that year. Beebe's promise to explore the depths of the ocean was sensational news in Bermuda, where the naturalist was already a local celebrity, so all eyes were on him and his little encampment on Nonsuch Island.

Beebe walked past the tidy dining room, heard no one stirring in the two small bedrooms beyond, and slipped onto the veranda that fronted the south side of the building. The air was clear, the sea calm, and he took in the view from the porch like a prince surveying his estate. In the dim new light he looked across the ragged shapes of the ancient Bermuda cedars that dominated the sparse vegetation of the island and the brushy understory of sage and goldenrod, the pink oleanders and red hibiscuses. He picked out Nonsuch Bay and South Beach, where he often swam in the evenings, and noted that the water there was as still as a lake with the barely perceptible soughing of the sea on the shore. The silver-yellow pulse from St. David's lighthouse was still visible beyond the beach, and he could just make out South Point three hundred meters away. Scanning from left to right, he saw Greene Island, where he had already spent many hours poking around the dens of the longtails, Gurnet Head Rock, and Castle Roads, where the shallow, sandy bottom afforded a safe anchorage in favorable northerlies and was a great spot to harvest conchs for chowder. Then there were Horne, Southampton, and Charles Islands, their limestone humps pocked with the first shadows of morning and invaded here and there by tufts of St. Augustine grass and lantana scrub. He ended his survey by marveling at the remains of the fort on Castle Island, the oldest evidence of the British presence in the New World, built in the early seventeenth century. Two shots fired from a cannon on the battlements of Castle Island had driven away a lurking Spanish ship in 1614, the only attack ever contemplated by an enemy against the English treasure of Bermuda.

Beebe absorbed the familiar panorama, then returned his gaze to the sea around Gurnet Head Rock a mile away to the south. He saw that it was calm offshore, too, with only a long, gentle swell warping the surface. Gurnet was forty feet high, and he had seen waves crashing over its top, even in the summer, but that morning he could barely make out the wash over the coral ledge guarding the north side closest to him. Beyond the rock, where the shallows of the fringe reef gave way precipitously to the abyss, the water was also calm, its deep blue surface just then picking up the first rays of sunlight at so low an angle that there was little reflection. The water looked like putty, with no streaks of white that would indicate wind. Soon the others would materialize from their tents and the two common sleeping rooms, observing as always the dictum that when the Director is awake and working, everyone is awake and working. Beebe heard the rattle and snort of the generator, which meant that Phil Crouch, who took care of the electrical system, was awake and tending to business. The clang of cooking pots announced Mrs. Tucker's arrival in the kitchen, and Beebe heard splashing from the cistern on the side of the main house—someone was hauling a pail of water for the breakfast chores. The chorus of morning sounds of an expedition camp was always reassuring to Beebe because it reminded him that he was exactly where he was born to be, preparing for another day in which nature might bless him with the thrill of discovery if he worked hard enough to earn it.

Beebe walked down the steps from the porch, looked at a thriving patch of cactus to his right, and went to tell Arthur Tucker, the island's caretaker, to hoist the green pennant up the flagpole on the north bluff. In St. George, the *Gladisfen*'s captain, Edward Millett, would see the signal from his veranda, walk down to the wharf, and charge the tug's boiler to build up steam for a day of work. An hour later, the craft would arrive off Nonsuch. Beebe returned to the kitchen and said good morning to Mrs. Tucker and Tee-Van, who was sitting at the dining room table looking sore and weary. Two days before on a dive together, Beebe's helmet had struck Tee-Van's chest when they both turned at the same time and he might have broken a rib or two. Beebe asked him if he was able to carry on, and Tee-Van told him he wouldn't miss the day if he were on his deathbed. Beebe smiled at the young man who carried so much weight on the expedition, gulped down his usual breakfast of Grape-Nuts and milk, and went to his room to pick up the canvas pouch he had packed the night before. In it were a plain lined notebook, several pencils, a bandanna, and his weathered copy of *The Depths of the Ocean: A General Account of the Modern Science of Oceanography Based Largely on the Scientific Researches of the Steamer* Michael Sars *in the North Atlantic.* Published in 1912 by John Murray and Johan Hjort, it contained drawings and descriptions of specimens gathered by trawls, traps, and hooks; the 800-page book was indispensable to any practitioner of the new science of deep-ocean ecology.

Beebe walked along a path that took him through the cluster of tents on the ridge plateau for the members of the expedition who weren't billeted in the main building. Everyone was awake and finishing their morning chores, shaking out sleeping bags, tying up tent flaps, washing up at the cistern, and passing through Mrs. Tucker's kitchen for a quick breakfast. Tee-Van caught up with Beebe as he was talking with Jack Connery, the young photographer from Florida who would be coming with them on the *Gladisfen* that day, making sure his cameras were working. The others gravitated to the three of them and formed a knot of khaki and chatter. There were Arthur Hollis, who kept the aquarium pumps working and helped out aboard the *Gladisfen;* Llewellyn Miller and Else Bostelmann, the artists; Virginia Ziegler, who was doing the typing and filing for the expedition; biologist Howard Barnes, an expert on invertebrates; Margaret Elliott, who worked with Gloria Hollister preparing fish specimens; and four schoolboys named Cannon, Guernsey, Potter, and Jackson, who were serving as field assistants. Phil Crouch walked over from nursing the generator and transferring gasoline to the day tanks, wiping his hands on a rag, and Arthur Tucker was right behind him. Gloria Hollister ambled by with her terrier, Trumps, and as though on cue the others followed her down the steep, narrow path to the water.

The landing was just an indentation in a thirty-foot bluff on the northwest side of Nonsuch. During the first expedition it had made for some precarious transfers between boat and shore. Beebe had remarked about the danger to William E. Meyer, from whom he had chartered the *Gladisfen,* and Meyer got permission from the government to salvage the wreck of the navy supply ship *Sea Fern* from the bottom of St. George Harbour, had it towed out to Nonsuch, and sank it again at a right angle to the shore to make a sheltered cove out of what had been a dicey landing. As a bonus, the *Sea Fern* had five holds which Tee-Van turned into tanks for specimens and food fish.

Under a sky brightening now through the yellows and reds of a tropical sunrise, Beebe and his Crowd waited for the smoke from the *Gladisfen's* funnel to appear in the gap between St. David and St. George Islands. Beebe's twenty-six-foot power launch, the *Skink,* lay alongside the dock with two outboard skiffs, and Tee-Van and Jack Connery went aboard the launch to get her ready to ferry the party out to the *Gladisfen* when the tug arrived. Two hundred meters beyond the *Sea Fern* breakwater, a long dark hulk lay at anchor, seemingly suspended over the lapis-colored water, her bow facing the rising tide and a light breeze. It was the 130-by-23-foot *Ready,* a former Royal Navy water carrier that had seen service all over the Atlantic but had been stripped of her rigging and retired to duty as a barge. The *Ready* was sound, but barely so. Her heavy timbers were rotten above the waterline from decades of freshwater seeping through every hairline crack in her decks and bulwarks, and she was riddled below by teredos and the infinite colonies of other creatures that inhabit wooden vessels in the tropics. It was easy to pick out the two steam boilers and the stout fifty-foot mast with a boom angled forward, but the men and women on Nonsuch landing had to squint to see the tiny white ball on the *Ready's* foredeck.

Barton had delivered on his promise to bring Beebe a new diving sphere. The winter had been wrapped in a shroud of doubt after the collapse of the stock market, as businesses closed every day and life was utterly transformed by what was just then being called the Depression. In an economic crisis, Beebe knew that his dream of descending into the realm of the creatures that tantalized him from his specimen jars and nets was fading. But Barton had soldiered on, and for this Beebe was so grateful that he had awarded him a title on his staff, Associate in Charge of Deep Diving. Beebe had also taken the liberty of naming Barton's craft. In the year and a half since Barton had shown up in his office on a cold December morning in 1928, Beebe, Barton, and Butler had been referring to it as a sphere, tank, bell, diving machine, and even cylinder. Finally, when Beebe was in his lab working on specimens from the 1929 season, he wrote down the name of a deep-sea fish—*Bathytroctes*—and was struck with the idea of using the Greek prefix *bathy-,* which simply means "deep," combined with "sphere." Perfect, he thought. This craft that he hoped would become as significant to the history of human exploration as Magellan's flagship *Victoria* and the oceanographic research pioneer HMS *Challenger* had a name: Bathysphere.

And there it was, on the foredeck of the barge anchored off Beebe's Nonsuch Island. The white globe picked up the faint rays of morning sunlight and winked over the *Ready*'s dark hull like an object descending from the sky instead of one firmly settled on the deck. It was hard for the members of Beebe's staff to imagine that the Director and Barton were actually going to get into the tiny sphere and have themselves lowered into the depths of the ocean, but this was the day they would try.

From the barge, the two Bermudian caretakers who had been left aboard as an anchor watch waved gaily to the crowd gathering on the shore, no doubt glad to have managed the night without incident and to be no longer alone with their barge and its strange cargo. The men turned and pointed to the south end of St. David Island, where they, and then everyone on shore, saw the pigeon-gray plume of smoke etching a wrinkle in the dawn-blue sky. The *Gladisfen,* with Otis Barton aboard, rounded into view a mile away.

The Bathysphere had arrived from New York two weeks earlier in late May in the cargo hold of the *Queen of Bermuda* and had been unloaded in a sling onto the Darrell and Meyer wharf in St. George. Barton arrived with it and succeeded in rigging the *Ready* using a triple-sheave system running off the Lidgerwood winch from the *Arcturus,* which had more than enough power to dead-lift the new, lighter, two-and-one-quarter-ton Bathysphere off the deck and manage it and three tons of cable in the water. An oil-fired boiler mounted aft on the starboard side of the barge generated steam at 130 pounds per square inch to drive the big winch, and a second boiler on the port side powered a smaller winch to swing the heavy boom. The *Arcturus* winch had a drum with a ten-inch solid steel core and a width between the outer flanges of five feet, six inches, on which had been wound 3,200 feet of seven-eighths-inch Roebling cable. Barton had brought 3,500 feet, but the drum wouldn't hold all of it.

The cable ran from the winch forward fifty feet to a self-lubricating steel pulley eighteen inches in diameter eye-bolted to a twelve-by-twelve-inch deck timber, then back to an identical pulley at the foot of the mast also bolted to a deck timber, and finally through a third pulley lashed and bolted to the tip of the twelve-inch-diameter wooden boom. The boom was swung from the mast with a quarter-inch cable tailing through a double block back to the small winch mounted aft between the boilers. Barton had checked every element in his rigging system with John Butler in New York and with the *Gladisfen*'s captain, Edward Millett. This time it would work.

As the *Gladisfen* crossed Castle Harbour and closed on Nonsuch Island, Barton stood on her bow, bracing himself against the forestay. He was dressed like a garage mechanic in dark green trousers and shirt with the sleeves rolled to his elbows and a greasy black leather skullcap, the same clothes he wore when he was working on the *Ready*'s rigging with the crew at the shipyard. He also wore an air of confidence about him that was more than just relief that one of the most frightening winters of his life was finally over. Something had annealed in Barton as he and Butler had struggled to overcome their disappointment in the failure of the first sphere and the financial nightmare of the stock market crash. The income from Barton's investment portfolio had suddenly dwindled to a trickle; he had been forced to to dip into his capital to keep the work going, bringing stern warnings from his mother and his investment bankers. Another particularly dark moment had come when the December issue of Hugo Gernsbacher's *Science and Invention* magazine carried a story that a descent had already been made into the abyss, complete with illustrations. Gernsbacher was an immigrant from Luxembourg who published magazines including *Science and Invention* and *Amazing Sto-*

ries, which combined reports on new discoveries and science fiction, often without any hint as to which category a story belonged to.

Butler had lost his job at Cox & Stevens, since not many yachts were on the drawing boards after the crash, and Barton's diving sphere was the only job he took with him from the firm. When he saw the Gernsbacher story, Butler wrote to his young client in a panic, afraid that the mercurial Barton would come unhinged again and pull the plug on the adventure that Butler knew was now costing him much more than he could afford. Barton, however, dug in against that and all other uncertainty about completing the diving sphere, and wrote to Butler:

> I have seen the December copy of *Science and Invention.*
>
> It may be possible that this feat has been accomplished. Many of the statements and pictures in the article however are obvious fakes. This magazine is also not considered reputable.
>
> I shall show it to Beebe next week.
>
> Meantime, I hope you will write to Roebling and the Morse Dry Dock about the cable.

On the *Gladisfen's* bow, Barton savored the morning. Nonsuch Landing came into focus gradually, as though through increasingly powerful lenses, and he could see the crowd there beginning to buzz with purpose as the tug approached. Soon he was able to pick out individual figures: he saw Beebe, Tee-Van, Hollister, Connery, and Hollis climbing onto the *Skink* with Arthur Tucker, who went forward to the steering station under the canopy. The people left on shore applauded and waved as the *Skink* edged away from the dock and started for the barge.

Aboard the *Gladisfen* with Barton were the deckhands from St. George whom Beebe had rounded up to man the barge for the dive, assembled on the fantail waiting to step over to the *Ready* when they were alongside. Beebe and Barton had figured out that they needed nineteen people, plus the five crewmen who ran the *Gladisfen,* to safely launch and recover the Bathysphere. John Tee-Van would be the deck officer in charge of all topside operations. Gloria Hollister would serve as the liaison between the deck and the divers, relay communications to Tee-Van, and take notes on the scientific observations passed up the phone line. She would also be responsible for calling an emergency end to the first dive if she did not hear a voice from the Bathysphere every five seconds. There would be one man on duty at the *Arcturus* winch, another at the smaller boom winch, one steersman manning the *Ready*'s rudder, controlled by a wheel eight feet in diameter, two men to clamp the telephone and electric line to the lifting cable as the Bathysphere went down, nine deckhands to pay out and haul in the cable and make sure it wound correctly on the winch drum, one man to tend the generator sending power to the Bathysphere, one messenger to keep Captain Millett informed and make sure Tee-Van and Hollister were each aware of what the other was doing, and one man to read the meter wheel on the *Arcturus* winch and tie tapes onto the cable every hundred feet.

Most of the Bermudian sailors who signed up to work on the Bathysphere dives were descendants of the survivors of the many slave ships that wrecked on the reefs of the archipelago in the eighteenth and nineteenth centuries. A few were salts from navy and commercial ships who swallowed the anchor in Bermuda and made a living doing whatever was necessary around the waterfront. Beebe offered from two to five dollars a day, depending on skills, enough to guarantee a full crew anytime the weather was good enough to go to sea.

Captain Millett swung the *Gladisfen* in a downwind arc to approach the *Ready*, then eased his sixty-five-foot tug alongside the barge, where the night watchmen passed down tie-up lines. The crew climbed to the top of the tug's deckhouse to make the transfer to the higher barge, reaching back for tool bags and lunchboxes, which were handed across. Barton made his way aft along the rail to the fantail and watched the *Skink* glide up to the *Gladisfen*'s rope ladder, then he lent a hand as Hollister, Tee-Van, Connery, Hollis, and finally Beebe clambered aboard. Beebe and Barton shook hands and stood face to face inches apart for a long moment, both men radiating solemn resolve, until Barton finally broke into a smile. Beebe nodded, looked over Barton's shoulder at Captain Millett, who was leaning out of his wheelhouse window, and gave a thumbs-up signal, then he, Barton, Tee-Van, Hollister, Connery, and Hollis climbed over onto the *Ready*. A crewman on the *Gladisfen* handed up a two-inch hawser attached to the tug's stern winch, which was then brought forward and wrapped around the tow bit on the bow of the barge. Captain Millett rang down to his engine room for slow ahead, and eased the tug forward while the crew manhandled the barge's anchor from the bottom and swung it aboard.

The *Gladisfen* steamed forward, her crew took up the slack on the tow rope, and *Ready*'s enormous bulk settled in behind the tug. Everyone aboard the barge walked to the bow, where no one could resist the impulse to reach out and touch the warm, white surface of the Bathysphere resting on the deck at the end of its cable like an amulet from some civilization of giants. The hatch was open, its cover lying on the deck, and everyone took turns peering inside. A few of the sailors looked, laughed out loud, and shook their heads.

As the procession passed the northwest bluff of Nonsuch, Beebe stood back from the Bathysphere and put his hands on his hips, a gesture that the others immediately interpreted as a command to get to work. Tee-Van took seven of the deckhands over to the Okonite cable laid out in long coils along the port rail and again went over the plan for paying it out as the sphere descended. Once the sphere was in the water and in no danger of smashing into the hull, the boom would be swung inboard so the cable hung straight down beside the barge and within reach. Every hundred feet, the Okonite hose would be clamped to the lifting cable to prevent twisting, and as the Bathysphere was hoisted, the clamps would be removed.

Hollis and Connery went aft to the winches and traced the rigging with their eyes to make sure that all the sheaves were clear and that there were no twists or trouble spots. A tangled cable with the Bathysphere in the water could be as much of a disaster as completely losing it. After running through the rigging, they checked the gas tank and connections on the Kohler generator. Gloria Hollister went to her telephone station on the starboard side of the mast and opened the plain, five-by-seven-inch, cardboard-bound school notebook in which she would write down the details of each dive and every word Beebe and Barton said to her over the telephone. On a loose index card taped inside the front cover was typed:

LIGHT SIGNALS

I FLASH = O.K

3 FLASHES = PULL US UP.

Among her responsibilities, Hollister would keep an eye on a signal light wired into the main electrical circuit that could be activated from the Bathysphere as a backup in case the telephone failed.

Hollister's notebook already included an entry on the brief test run to forty feet at the bottom of St. George Harbour on May 27 with Beebe and Barton aboard. They hadn't planned to make the shallow dive, but Beebe had coaxed Barton into the Bathysphere to test the oxygen system and decided to seal the hatch and give the crew the experience of handling the sphere before they were on the open sea. Tee-Van had only hand-tightened the hatch bolts, thinking that the water pressure was nothing to worry about, just two atmospheres greater than at the surface. The dive had lasted ten minutes and Beebe and Barton had emerged safely after seeing nothing but muddy water. Hollister had noted that the Director and Mr. Barton had no trouble breathing, but seemed slightly giddy from the oxygen.

Hollister had also already made notes on the test of the empty Bathysphere to a depth of two thousand feet on June 3, six and one-quarter miles south of Nonsuch, precisely at 32°14'30" North, 64°40' West. It had been a disaster. As the sphere had descended at a rate of one hundred feet every two minutes, the lifting cable had slowly twisted and wound the pliable rubber Okonite line around itself. On the ascent, the lifting cable had not untwisted in the opposite direction, so the Okonite line had been draped in forty-five loops over the sphere when it had been brought aboard the *Ready*. The tangle had looked hopeless, and Beebe and Barton had thought they had failed again. Back at the wharf, though, Tee-Van had figured out that when they wound the Roebling cable from its shipping spool onto the winch drum, it had twisted very slightly and then untwisted as the Bathysphere descended, winding the Okonite line around it. Tee-Van and a crew of dock workers had removed the cable, laid it out on the wharf, and rewound it free of the deadly twist back onto the winch.

Now Hollister entered June 6 as the date of dive number 3, which Beebe had told her would be another test on the open ocean. If the Bathysphere came back intact and dry, and if the cable didn't tangle, Beebe and Barton would attempt the first manned descent later that same day.

Seven

THE ABYSS

Had Heebie-Jeebies
for Beebe's Bathysphere

Stephen Sondheim

astle Island passed astern and the *Gladisfen* and the *Ready* entered the open Atlantic. The wind was light and variable, but a storm somewhere to the southeast announced itself as a gentle beam swell that set the barge's old hull to creaking. The tug made only three knots, about as fast as a person can walk, and Beebe dead-reckoned his position off that speed and by triangulating with the St. David's and Brangman's Head lighthouses. He was heading in the general direction of his cylinder, but nothing precise about the surface would tell him when he had arrived in the right place. Beebe knew, though, that five miles offshore the steep flank of the mother volcano that had created the Bermudas would be behind them and the bottom at least two miles beneath the *Ready*'s keel. He wasn't quite sure how deep the first manned dive would be, but definitely not the half mile that the newspapers were touting as Beebe's goal, just under the maximum depth possible with the 3,200 feet of cable if they allowed for the rigging and a safety margin of cable left on the drum. The end of the cable tearing free from the winch and sending them to the bottom was one of the many nightmares Beebe and Barton worked to banish as they prepared for their first dive.

Suffocating inside the Bathysphere was another. For most of the hour it took the tug and barge to reach deep water, Beebe and Barton went over the equipment that would enable them to breathe. For its design, John Butler had gone to Dr. Alvin Barach, an expert on pressurized gas in New York who invented a special valve and a gauge to measure oxygen flow. The Bathysphere was fitted with two removable tanks, each seventeen inches long and four and a half inches in diameter, that clamped into horizontal brackets opposite each other on the lower quadrant of the sphere. Each tank contained three hundred liters of oxygen compressed under 1,800 pounds of pressure; together, they could sustain life in two people for up to six hours. The adjustable valve and gauge were fitted to one tank and could be switched over to the second if necessary, discharging oxygen into the Bathysphere at two liters per minute. Beebe and Barton were not sure of the exact rate, though, and would experiment using themselves as guinea pigs. If they had trouble drawing breath or felt drowsy, they would increase the flow; if they felt drunk from too much oxygen in their blood, they would turn it down.

Racks for holding open metal trays similar to bathroom shelves were welded above each oxygen tank, one tray to contain soda lime to absorb the carbon dioxide in the air, the other calcium chloride to absorb moisture. The soda lime was Wilson Sodasorb, as fine as silica sand, and the calcium chloride was roughly the consistency of coarse beach sand. Dr. Barach told Barton that he would need one pound of each chemical per person per hour, so two extra one-pound bags of each would be stowed

inside the cramped Bathysphere. Barton would circulate air over the trays by waving a palm leaf fan he bought at the market in St. George.

Beebe looked up at the position of the two lighthouses and saw that they had a few minutes to go before launch, so he and Barton checked the filaments that would connect them to the world above during the descent. The lifting cable had been threaded through a heavy fitting called a clevis and then separated into individual strands of wire between which molten steel had been poured to form a solid knot after it hardened. The clevis was pinned to the center hole in a three-hole lifting lug by a one-and-a-quarter-inch stainless steel bolt and secured with a one-quarter-inch steel cotter pin.

The telephone and electrical cable entered the sphere through the stuffing box, the outside of which was a two-and-a-half-inch steel fitting threaded through a reinforcing plate just to the side of the lifting lug. The fitting had been drilled through the center to slightly more than the one-inch diameter of the power cable that passed through it. The stuffing box was completed by a brass fitting inside that could be tightened with a wrench. The Okonite cable split into two lines, one leading to a headset similar to those worn by telephone operators with the curved horn of a microphone fitted to a wire support that hung around the neck. The other line went to the housing for the General Electric searchlight, which was mounted on the wall of the sphere and aimed out the porthole on the right side. The center hole was the viewing port and the left hole was closed by a steel plug.

When the windows were installed at Watson-Stillman, two of them cracked, one when the housing ring was tightened unevenly and the other during a pressure test at just 250 pounds per square inch. Barton was convinced that the glass that failed under pressure had done so because of a defect and the other broke because of human error. The two windows that survived installation had passed pressure tests to 1,500 pounds per square inch. They were set in place over a seal of white lead paste, and secured by a round steel collar fastened by ten one-inch bolts on threaded brass studs, leaving a viewing area six inches in diameter.

Beebe and Barton moved deliberately with no idle chatter as they inspected their craft. Then the two men stood on the *Ready*'s foredeck with the Bathysphere on their left while Jack Connery took a photograph. Posing for a snapshot at a ceremonial moment was familiar territory for Beebe, who appeared relaxed though distracted, but Barton looked worried as he stared directly at the camera with his lips and his brow creased. After Connery tripped the shutter and went back to work with Tee-Van, Beebe reached across the Bathysphere, put his hand on Barton's shoulder, and smiled. Beebe walked away from Barton and the Bathysphere and over to the port rail, where he looked down at the loops of the Okonite cable, and then stared out at the sea which spread endlessly to the south. The surface shimmered with the yellow patina of new sunlight. He realized that it would never again look the same to him after that day. A few weeks earlier, Beebe's friend Percy Crosby had drawn a cartoon for him in which two ragamuffin little boys are standing at the edge of the water on a beach. One of them says, "Gee, the ocean's big, ain't it," to which the other replies, "Yeah, an' ya' mustn't forget, that's only the top of it."

The sun was two hand-widths above the horizon when Beebe looked at the position of the lighthouses on shore and decided he was over his cylinder. He smiled to himself as he thought that choosing an exact spot on the sea was akin to deciding on the exact location of the North Pole. Some Arctic explorer, he remembered, said it was just a matter of getting close and sitting down anywhere on the ice. Beebe waved a signal to a deckhand on the *Gladisfen*'s stern two hundred feet ahead of the *Ready*'s bow, who ran forward to tell Captain Millett, who rang dead slow to his engine room. The tug hove to against the light wind and swell and settled onto a southeast heading. Millett's crew winched the *Ready* in from her more distant towing position until the barge trailed about fifty feet astern. Because it was too deep to anchor, the tug held station under power, and Millet pinpointed his position on the Admiralty chart at 32°11' North, 64°39' West.

Aboard the *Ready*, the expedition party wasted no time. Tee-Van closed the hatch cover, pounded the bolts tight, and was ready to launch the empty Bathysphere for the test to two thousand feet. Beebe gave a thumbs-up, Hollis opened a valve sending steam to the winch, and the white globe seemed to leap off the deck and swing through an arc of a few degrees with the gentle roll of the barge. As they had practiced twice before, five of the Bermudian seamen formed a line to pay out the Okonite cable, two stood by the rail to attach it to the lifting cable with a brass clamp every one hundred feet, and two others handled a rigging line and the small winch to swing the boom out over the water. On Tee-Van's command to lower away, the Bathysphere splashed into the sea and the white globe shimmering in the purple water quickly became a speck and then nothing. The steady thrum and creak of the cable through its sheaves, the grinding of the winch, and the scrambling of the deckhands feeding out the Okonite cable were the only evidence that the Bathysphere was descending.

The tension on deck was palpable. During the first unmanned test three days before, the disaster of the twisted cable almost ended the great adventure before it had really begun. Barton had been crestfallen, but Beebe reassured him with the paternal certainty that was so much a part of his personality. "Remember, Otis, this has never been done before," he said. "You can't expect things to lie down for you." With the sphere in the water again, Barton and Beebe paced from winch to foredeck, but let Tee-Van and the others do their jobs because they would not be on deck to supervise when it really mattered. An hour after it vanished into the sea, the Bathysphere was safely back on deck. The lifting cable had not twisted, the Okonite hose was perfectly retrieved and again flaked into coils on the deck, the windows were intact, and nothing about the outside of the Bathysphere looked any the worse for wear from its trip into the abyss. Beebe and Barton removed the hatch cover themselves and were relieved to see that less than a quart of water had collected in the bottom of the sphere, most likely from the predicted leaking of the stuffing box.

Beebe and Barton dried and cleaned the inside of the sphere, loaded in the oxygen tanks and chemicals, and paused for a few minutes to absorb the magnitude of what was about to happen. Mrs. Tucker had sent lunches along with them, but no one wanted to eat. Suddenly there was nothing left to do but go. Beebe tried to think of something ceremonial to say, but even this lifelong devotee of Kipling and his heroic poetry was struck dumb. He looked around at the sea and sky, the boats and his friends, but he still couldn't think of anything pithy or monumental. So he said nothing.

Beebe and Barton climbed in over the hatch bolts, a painful process of slipping headlong into the sphere, hands, arms, head, then shoulders one at a time, then hips, which were easy for the slimmer Beebe, who entered first, but which required squirming for the heavier Barton. The hatch was only fourteen inches in diameter, designed by John Butler to be the absolute minimum size to reduce the compromise to the strength of the sphere. Inside, the men crouched because there was no other way to fit.

They tested several positions: the best seemed to be one of them sitting with his knees drawn up and the other kneeling with his butt back on his heels. At this point, Beebe asked Tee-Van to have someone fetch him a cushion, but after a few minutes of searching, none was found.

Barton opened the valve on the oxygen tank, heard the reassuring hiss of the gas, and tested his fan to see that he could reach over the chemical trays. As they struggled to fit themselves into a space that would have been tight for even one of them, Beebe quipped that he had no idea there could be so much room in a four-and-a-half-foot sphere. Barton settled against the wall on the side of the main hatch and put on the telephone headset, identical to the one Gloria Hollister wore on deck. He said, "Testing, testing, one, two, three," and Hollister answered, "I read you loud and clear, Mr. Barton."

Under the noon sun, the Bathysphere was quickly becoming an oven, and sweat poured off the two men. In the dim interior light Beebe and Barton looked each other right in the eyes; both would remember that the fear they expected to see was not there. Then Beebe simply nodded to Tee-Van, who was leaning over and looking through the opening. Tee-Van nodded back to Beebe, stood up, and began the process of sealing the hatch. The 400-pound, three-inch-thick steel cover was suspended from the boom on a half-inch line run through block and tackle and manhandled by two deckhands. Tee-Van guided it into place over a soft copper gasket and hand-tightened the ten bolts over the threaded brass studs.

One of the last refinements John Butler had made was to bore and thread a four-inch hole in the center of the circular hatch cover into which he screwed a removable wing bolt. The idea was that in an emergency, especially one involving the oxygen system, the wing bolt could be removed in a matter of seconds, while the hatch, secured by ten tightened bolts, took five minutes or more to open. Once the hatch cover was on, Beebe put his hand through the center hole to shake with Tee-Van, then Barton did the same. Tee-Van screwed in the wing bolt, and he and another crewman dogged down the ten main bolts with wrenches, working opposite studs to seat the hatch cover evenly on its gasket. Tee-Van tightened the bolts with hammer blows, each ringing the Bathysphere like a cathedral bell. Inside, Beebe and Barton cringed as the cacophony went on for ten minutes that seemed like an hour. Beebe was sure the hammering would crack the windows. He shouted across Barton into the headset, but couldn't hear Hollister's reply, and the clanging continued until all the bolts were tight. Then utter silence settled in.

The dreadful consequence of failure was on everyone's mind that day off Bermuda, but no one admitted to it, and the work of launching the Bathysphere proceeded with the efficiency of a freightyard. Hollis again turned a valve to direct a head of steam to the winch, Crouch brought the generator up to speed, and Tee-Van looked for a signal from Captain Millett up on the *Gladisfen*'s bridge. The captain was waiting for a propitious set of swells to swing the sphere over the side. The light wind of the morning had built into the thermal breeze of early afternoon, and the ocean had responded with a chop that tossed the tug and barge around enough to worry Millett. He would have a mass of two and a half tons swinging in the air at the end of a cable, and a good hit could put a hole in the *Ready*'s rotten flank. If the barge sank, the Bathysphere, Beebe, and Barton would go down with it.

At exactly one o'clock, Captain Millett waved down to Tee-Van to engage the winch, the cable tightened, and the sphere rose six feet over the deck, dangling in midair and revolving slowly. Beebe and Barton felt the Bathysphere swing free and through the portholes saw the horizon, the sea, and the scrambling crew pass by in a panorama of all they were leaving behind. Two of the deckhands steadied the sphere and others wrestled the boom with hand lines and swung it across the bulwark, where it hung, poised over the ocean. Inside, the motion was amplified, and Beebe, fearing a crash that would break windows or sink the barge, blurted out a curse. Barton said to him, "Miss Hollister wants to know why the Director is swearing so." Clearly, the acoustics of the sphere would send every word they uttered through the telephone line no matter who was wearing the headset. Beebe continued to rant until Hollister told them that they were then hanging twenty-five feet from the side of the barge and in no danger of hitting anything. He apologized profusely to Hollister, rattling on until the Atlantic swallowed the Bathysphere and he turned his attention to the ocean.

Beebe and Barton were helmet divers, so the sea in the first fifty feet of the descent was familiar to them. They saw the beard of sea grass and barnacles on the *Ready*'s bottom and were delighted that their vision was so clear through the glass. They had no way of knowing how deep they were from inside the sphere, so Gloria Hollister called out their depth from a measuring gauge on the winch. The *Ready*'s keel passed slowly out of view above them and their last visible link with the upper world was gone. From now on, Beebe and Barton had to depend on distant spoken words for knowledge of their depth, their speed, the weather, the wind, or anything else having to do with the bright, airy world on the surface.

Hollister called down, "One hundred feet," as the sphere jerked violently and stopped. Barton's eyes widened, but Beebe reminded him that the stop was planned so that the crew could attach a clamp to fix the Okonite power and phone line to the lifting cable. They began to descend again, and Beebe patiently made notes about Commander Edward Ellsberg and his gallant, failed attempt to rescue the crew trapped in the navy submarine *S-51* at a depth of 132 feet in 1925.

Otis Barton fanned the chemical trays and both men fought back waves of fear. "Breathe, Otis, breathe," Beebe said, as much to himself as the young man he saw trembling beside him in the dim light. At two hundred feet, they jerked to a stop again and for a long minute in the darkness of the Bathysphere they swung through the water in a gentle arc, keeping time with the *Ready*'s roll on the surface. Finally, with a lurch, they started down again.

"Three hundred feet," Hollister said, and just as the Bathysphere shuddered to a halt for a cable clamp, Barton cried out, "We're leaking." Beebe turned on his flashlight and the men watched a trickle from the lower right-hand arc of the hatch feeding an accumulation of about a pint of water at their feet. The water followed the curve of the sphere in a weak stream that didn't seem to be increasing. Beebe studied it, and after a glance at Barton, who shrugged his shoulders, ordered the descent to continue. Though the flow wasn't increasing, they kept an eye on the leak as if it was a coiled viper ready to strike.

Beebe and Barton had been sealed in the sphere for twenty minutes, in the water for five, and already it seemed like an eternity. In three more minutes they hit four hundred feet. The stop for a clamp was no less frightening than it had been the first time—a startling lurch like a dark elevator shuddering to a halt between floors. Beebe tried to focus on the sea outside the porthole but with little success, sweeping his gaze back and forth from the window to the leaking hatch. Every five seconds, Barton muttered "Fine, fine," or "Keep going," or simply "Okay," and the inside of the Bathysphere grew colder and darker with every foot it descended.

When Hollister called out six hundred feet and the sphere jerked to a stop, a shower of sparks erupted from the electrical line where it entered the searchlight. Bright yellow flashes illuminated the sphere, and Barton instinctively reached past Beebe's head to grab at their source. He knew that there wasn't enough free oxygen in their atmosphere to ignite, but if a spark reached the oxygen valve itself it would turn the Bathysphere into a roaring furnace. The problem was a loose connection to the searchlight box. Barton wiggled the fitting, and darkness returned, now tinged with the ozone stench of free electricity. Both men were trembling uncontrollably but passed the word up to Hollister to continue the descent.

At seven hundred feet, the hatch and the window seals creaked under the immense pressure, but they were holding. Beebe ordered a halt for observation and they remained there for five minutes. He was finally able to sit calmly at the porthole with Barton looking over his shoulder. Outside, the color of the sea had deepened to a dark translucent blue. Beebe thought of where he was, this place where no human being had ever been, and finally found something ceremonial to say.

"Ever since the beginnings of history, when first the Phoenicians dared to sail the open sea, thousands upon thousands of human beings had reached the depth at which we were now suspended, and had passed on to lower levels," he intoned to Hollister. "But all of these were dead, drowned victims of war, tempest, or other Acts of God. We are the first living men to look out at the strange illumination."

And it was stranger than any imagination could have conceived, an indefinable blue quite unlike anything in the upper world. Beebe flicked on the searchlight; its beam was the faintest yellow he had ever seen. He let it soak into his eyes, and switched off the beam. Though faint traces of visible sunlight reach to a depth of almost two thousand feet on a bright day, Beebe and Barton were the first men to enter the realm in which most of the waves (or photons) of light generated by the sun have lost their power to penetrate the human retina, leaving only the violet-blue they were seeing. The sun's light and colors simply drop away in order of the position of their wavelengths in the spectrum, beginning with reds at about 50 feet, then yellows at 150, and greens at 300. Finally only the faintest hint of purplish blue remains at 700 feet.

They continued the descent in silence except for a word or a grunt every five seconds to let Hollister know they were still alive. The leak was no worse, the fire was out, and they were breathing fine, if a little light-headed from the oxygen. But at eight hundred feet a stark premonition of disaster invaded Beebe's mind and he abruptly called a halt to the descent. He had been thinking about going to one thousand feet, a nice round number, but some mental warning that he trusted spelled the end of that first trip. That settled, they hung at 803 feet, Barton tended to his air bottles, chemical trays, and the palm fan, and Beebe looked out the window.

He saw creatures flashing through the beam of the searchlight that defied even his vivid imagination. There were fish with gaping jaws that exposed forests of needlelike teeth, schools of squid swimming together as though synchronized to the beat of a choreographer, and pulsating chains of jellies many meters long. Beebe had to force himself to breathe. No one had ever seen anything even remotely like the animals of the abyss in their natural setting. With the searchlight off, Beebe and Barton saw galaxies of bioluminescent streaks flash across the windows in a sub-marine pyrotechnic display that was beyond belief.

Beebe wrote very little in his notebook about life outside the sphere on this first dive, but good records could come later. Controlling fear, watching the leaking hatch, fixing the short circuit, marveling at the transformation of light, and adjusting to the other realities of the Bathysphere at depth were enough. The first descent was a pure test of survival. After five minutes at eight hundred feet, Beebe told Hollister to begin the ascent. "Are you okay?" she asked, her voice rising to a shrill register above its normal calm tone. "Yes," Beebe told her. "It's just time to come up. Sounds like you are doing the hardest part of this job up there."

The return to the surface took an uneventful twenty-one minutes, the entire dive from deck to deck just under an hour. As they rose, elation built up in both men as though the brightening water was banishing fear with every inch that they drew nearer to the light. The crew hoisted the Bathysphere back aboard the *Ready* while Beebe and Barton sat patiently peering out the portholes until the wing bolt was unscrewed. Finally, with a hiss of compressed air escaping from the sphere, they were back in the world of the sun. Beebe stuck his hand out the center hole and Tee-Van, then Hollister, shook the ends of his fingers. Then Barton did the same. The clanging of the hammer loosening the bolts was just as loud as it had been when they were sealed into the Bathysphere, but infinitely easier to bear this time. Finally, the hatch cracked open, fresh air and voices rushed in to replace the ozone and sweat, and it was over.

Following naval protocol that the senior officer be the last to leave a craft at sea, Barton climbed out first, then Beebe. On deck, they wobbled on cramped legs while Connery took their picture and the screeching whistles of the winch boilers and the deeper-toned siren of the *Gladisfen* saluted them. On Nonsuch Island, the rest of the Crowd at the field station saw the puffs of steam from the whistles and knew that William Beebe and Otis Barton were alive.

While the others celebrated, Gloria Hollister dutifully computed times, pressures, and temperatures and completed her entries in the Log of the Bathysphere.

> *Dive # 3 W.B. and O.B.*
> *Traveling time in motion:*
> > *Down—24 minutes, 100 feet every 3 minutes.*
> > *Up—10 minutes, 100 feet every 1¼ minutes.*
> > *Total average time per 100 feet: 2.17 minutes.*
> > *Total average time per 1 foot: 1.3 seconds.*
> *Bell in sight for 138 feet or 42 meters.*
> *Temperature: On deck 87°, 600 feet 78°, 800 feet 76°.*
> *Pressure: 360 lbs. per square inch.*
> *803 feet, 134 fathoms, 245 meters.*

Eight

INFINITESIMAL ATOMS

*She tried to fancy what the flame of a candle
looks like after the candle is blown out.*

Lewis Carroll, *Through the Looking-Glass*

For four days after the first dive, a summer storm threatened to become a hurricane and kept the *Ready* at anchor in the lee of Nonsuch, where the old barge bobbed and swayed in the chop with the Bathysphere lashed down on deck. Everyone on the island was distracted and anxious to get on with the diving, which had been transformed from a flight of the imagination into stunning reality. Beebe and Barton had emerged giddy with slight oxygen jags and Beebe's chronic sinus condition flared up, but otherwise they were in good shape. The Crowd from Nonsuch celebrated the new world record of precisely 803 feet with dinner and dancing at the St. George Hotel on the night of June 6, but the party was tame by Beebe's standards. The first dive was just the beginning, and he was leery of premature declarations of success. It was an early night, and everybody but Barton piled into the *Skink* and went back to the island. The next morning they dove into a frenzy of work that was an antidote to worrying about the weather. Beebe buried himself in his notes, already planning his book on the Bathysphere, and Jack Connery developed his photographic plates. From Beebe's description of the shrinking of the spectrum during his descent and a textbook on the physics of light, Else Bostelmann painted a watercolor spectroscope. It was an inverted cone that showed the visible spectrum from red on top, through orange, yellow, green, blue, and finally purple on the bottom. With it, Beebe could more precisely record the deterioration of the spectrum by simply holding the spectroscope near the window and noting the depth at which each color turns to black. The observation of light in a descent into the abyss was not only interesting, but it explained the coloration of fish, shrimp, and other creatures in relation to one another and the particular strata of light in which they lived. Beebe was determined to prove that the Bathysphere expedition was not a stunt for setting records but a revolutionary tool that would transform the young science of oceanography.

Barton stayed in St. George and made trips across the harbor to the *Ready,* where Beebe joined him for several hours each day. They worked on getting a better seal on the hatch with more white lead paste and inspected the windows, detecting what looked like a small crack near the outer rim of one of them. It didn't look serious, maybe just a flaw in the quartz, but it meant that they had to run a deep test dive before another manned descent. At Beebe's suggestion, Barton painted the interior of the sphere around the windows black to reduce glare from the searchlight and make observation easier.

On June 10, the wind finally died, the sea lay down enough for a safe launch of the Bathysphere, and Beebe raised the signal flag on Nonsuch bluff to bring the *Gladisfen* across Castle Harbour. By ten, the empty Bathysphere was on the way down to two thousand feet to make sure the

window wasn't leaking and to give the crew another drill. They left it in the water for two hours, half of that at maximum depth. When it was back on deck the window hadn't leaked but somehow three feet of the Okonite hose had been forced through the stuffing box and into the sphere. Barton wrote off the problem to a loose connection, which he assured Beebe he would have been able to tighten during a manned descent. They knew that the invisible demon of water pressure would be a dark companion on every dive, and the hose forced into their Bathysphere was a stark reminder of its presence.

Beebe and Barton adjusted the hose, tightened the stuffing box, recharged the chemicals in the trays, and clamped the oxygen bottles in place. Then they climbed into the Bathysphere, in what had already become a ritual of grunting, groaning, and mock bravado as they pulled themselves over the hatch bolts. This time, they didn't forget cushions, Barton made sure he had his lucky greasy skullcap on just right, and Tee-Van apologized in advance for the banging of the hammer that would begin when he dogged down the hatch. Gloria Hollister wired a live lobster to the housing of the middle window to attract fish, and Tee-Van attached the flags of the Explorers Club and the New York Zoological Society. The Bathysphere looked like a well-decorated ornament as it went into the sea at 12:45 with Beebe and Barton aboard.

At one hundred feet, Beebe told Hollister that the water was full of small jellyfish, and she made a note. A minute later, he said he saw two jacklike fish swimming past, and one good-sized Aurelia jelly, then two long strings of larva colonies called salps. Just past two hundred feet, Beebe was telling Hollister that he felt a definite sensation of weightlessness as the sphere descended when the phone line crackled and went dead. "My God. The phone is broken," Barton said. Beebe's mind flashed to the leak and the short circuit of their first dive, and realized that while those were terrifying, this was worse. The sudden absence of the only human voice connecting them to the familiar world drove home their isolation. Barton flashed the searchlight three times. It was wired to also light up a signal bulb on deck when the circuit was energized, so Hollister saw the emergency signal. Seconds later, they felt the Bathysphere bounce to a stop as though hitting a cushion of foam, and then they were pressed to the floor as they shot to the surface. Typically, some part of Beebe's naturalist brain remained on duty even in the crisis, and he counted twenty-four jellies on the way up. The sphere swung out of the water and onto the deck in two minutes, and Barton rapped on the side with the wrench to let them know they were alive. Beebe looked out at his crew and saw that Tee-Van, Hollister, and the others were ashen and obviously frightened, so he pressed his face to the center window and gave them a reassuring grin. The crew's expressions remained strained, and Beebe knew that their terror over the past five minutes had far exceeded his own. Then the wing bolt was off, and the hatch cover, and the second attempt at a deep dive was a failure at just 250 feet.

The weather continued to cooperate, but the ride out to the sea the next morning was anything but festive. The flicker of confidence that had been kindled by the first successful descent to 803 feet had been doused by the brief but agonizing moments when the phone went dead. A broken wire had caused the trouble in the telephone, and Tee-Van located it in the fifty feet of hose nearest the sphere, so the repair was a simple matter of cutting it off and rewiring the line. The mood was also dampened because a fire had broken out aboard the *Ready* during the night, probably started by oil ignited in one of the still-hot boilers as the barge lay at anchor. The blaze had burned the deck and part of the aft gunwale, but the winch, Okonite cable, and Bathysphere were untouched, which Beebe interpreted as good luck. He had trouble selling his optimism to Barton, though, who tended toward pessimism. So far, in Barton's mind, the Bathysphere expedition had come near to disaster five times—the tangled cable, the leak, the short circuit, the dead phone line, and the fire on the *Ready*—and he was brooding.

Beebe, though, was ready to get down to business. Every foot beyond 803 was not only a new world depth record, but unexplored territory, and he wasn't about to let a few minor problems slow him down. As the procession of the tug and the charred barge steamed through Castle Roads, Captain Millett yelled back to Beebe to tell him that a strong current flowing out of the harbor would make conditions in the usual cylinder dangerous, so diving closer to shore, where the current of the ocean back-eddied, would be safer. Beebe agreed, checked a chart, and found a spot five miles offshore with depths of 4,500 feet that would do just fine for the 1,500-foot dive he wanted to make that day. The 803-foot descent had given Beebe acceptable odds for survival, but he was through being a test pilot. It was time for science.

After the aborted 250-foot descent, Beebe and Barton had made a few more changes to their craft. The lifting lug on top of the Bathysphere had three holes running fore and aft, and on the first two dives, they had fitted the clevis bolt through the center hole. Now they moved it to the rear hole so the windows would be aimed at a slightly downward angle when the sphere was suspended, giving them a view of deeper water. They also refined the interior to include a small shelf wired to one of the chemical trays for Beebe's pencils, Bostelmann's spectroscope, and two specimens of scarlet shrimp against which to also compare the light at various depths. They stored their spare palm leaf fans and the stuffing box wrenches on the bottom of the sphere behind them.

Tee-Van had come up with the idea of laying burlap sacks over the hatch bolts, and Beebe congratulated him with a comedic bow and shuffle as he boarded the Bathysphere just before ten o'clock on the morning of June 11. Barton tumbled in behind him, and they endured the ritual of slamming home the hatch bolts and the wing nut. Just before Beebe gave the order to launch, Barton discovered that he didn't have his lucky skullcap. Tee-Van hammered open the wing bolt and the *Ready* was thoroughly searched for five minutes, at which point Barton discovered that he was sitting on the hat. Tee-Van reset the wing bolt, checked the lines that secured the flags of the New York Zoological Society and the Explorers Club and a cheesecloth sack holding a rotted squid under the center window, and signaled the winch man to haul away.

Four minutes later, the Bathysphere was in the water. A few feet under the surface, Beebe seemed euphoric and waxed poetic through the phone line to Gloria Hollister. He told her the towline to the *Gladisfen* looked like a sea serpent, and the boundary between air and water above him like a slowly waving, green canopy, quilted everywhere with deep, pale puckers. The sunlight sifted down in long, oblique rays as if through some unearthly, beautiful cathedral window. The water was so clear, he said, that he could see the *Gladisfen*'s distant keel decorated with seaweed like mistletoe hanging from a chandelier. Beebe asked Tee-Van to lower the Bathysphere slowly at no more than an average of a foot per second so he could bring his full observational powers to bear on what was beyond the window.

At fifty feet, Beebe glanced at the scarlet shrimp on his pencil tray and was startled to see that they were a deep, velvety black. He opened his copy of *Depths of the Ocean* to a plate of bright red shrimp: they, too, were black as coal. As the Bathysphere descended, he compared the light outside the sphere to Else Bostelmann's spectroscope, scribbled in his notebook, and talked to Hollister, who made her own notes. "Just beneath the surface red one-half normal width," she wrote. "At 20 feet, only a thread of red. At 50 orange dominant."

Beebe was mesmerized by the physics of light as they descended and kept up a running banter to Hollister, who logged the spectral transformations and the creatures passing by. At one hundred feet a cloud of thimble jellies vibrated twenty feet from the window, animals that until then were believed to be surface dwellers. Since he knew their actual size, they were a gauge of comparison for estimating distance, size, and speed of unknown organisms and proof that the quartz glass was not distorting size in his vision.

In the short time they had actually spent inside the Bathysphere, Beebe and Barton had become acutely sensitive to their little artificial environment. At two hundred feet, the oxygen tank was working perfectly, the palm leaf fan was keeping the air comfortable, and they already sensed the drop in temperature. Their thermometer read 78 degrees, a loss of ten degrees from the midday heat on deck. Suddenly, a six-inch fish appeared in the window, nosed the bag of squid, and then held its position near the glass. Something about it seemed familiar, yet it was strange. It looked a lot like a pilot fish, *Naucrates ductor,* but it was pure white with eight wide, black, upright bands. Beebe realized that at two hundred feet a pilot fish would not be the color it was at the surface. Its reds had gone to black, its grays to white, so it was like the ghost of a pilot fish.

As they descended through three hundred feet, Beebe called up sightings of more pilot fish; a small fish, *Psenes,* which has no common name; and long strings called siphonophores, a colonial jelly, as lovely as the finest lace. He saw small vibrating motes pass in clouds and it took him a moment to figure out that they were pteropods, or flying snails, each of which lived within a delicate tissue shell and flew through life in the sea with a pair of flapping, fleshy wings. "At 300 feet," Beebe dictated through the phone, "the whole spectrum is dimmed, the yellow almost gone and the blue appreciably narrowed. At 350 spectrum 50 percent blue violet, 25 percent green, and an equal amount of colorless pale light."

At four hundred feet, Beebe saw the first real deep-sea fish, *Cyclothones* or round-mouths, lantern fish, and bronze eels. Of the thousands of these fish he had netted, he had never seen one alive until then, and instead of having only a few scales left from the torment in the nets, these animals were ablaze with their full armor of iridescence. As he watched one of the lantern fish, he saw the flash of its light organs. Then a small puffer came into view, another species thought only to inhabit the shallows, and Beebe mused that he in his white Bathysphere must appear much more out of place to the little fish than vice versa. "At 450 feet, no green remains, only violet, and blue too faint for naming."

At five hundred feet with the searchlight on, a shape over two feet long ghosted past, and Beebe saw large, dark forms hovering just past the edge of the darkness that reminded him of wolves lurking beyond campfire light. Squid balanced in midwater, and a four-inch fish nosed one of the baited hooks near the center window. They could see the vertebrae and body organs through its transparent body and he identified it as a juvenile big-eyed snapper. Past six hundred feet, a dozen fish Beebe couldn't name swam by with their noses toward the surface, and shrimps and snails drifted past like flakes in an unearthly blizzard. A large transparent jellyfish bumped the window, and Beebe saw that its stomach was filled with a glowing mass of food. "At 800 feet, no color visible but pale grayish-white in the violet-blue area."

They had been in the water for twenty-five minutes when they passed eight hundred feet. Outside, Beebe again saw only the deepest, blackest blue imaginable, the unearthly color that thrilled him and Barton beyond their wildest expectations on their first dive. "Look, Otis," Beebe said. "No human being has ever seen that except us and we've seen it twice."

Beebe suffered no fatal premonitions this time, so they eclipsed the record of their first dive and continued down. The value of that achievement faded quickly when Beebe made the first human sighting of a school of living hatchetfish, *Argyropelecus,* flashing their silver sides in a conga line in front of the window. He tugged Barton over to his window to verify the identification. The beam of the searchlight was on, barely visible in the brilliance of the deep blue water, but they could see fish swimming in and out of it. Beebe chattered nonstop up the line to Hollister. He reported two black fish eight inches long, rat-tailed *Idiacanthus;* two long, silver, eellike fish, *Serrivomer;* myctophids with headlights; pteropods; more eels; and a big *Argyropelecus* that, head-on, looked like a worm.

Two minutes later they were at nine hundred feet, passing through a mist of shrimp and flapping snails. Fear struck at a thousand feet when something black and snakelike swung around and banged on the windows. They turned on the searchlight, which was more effective now that the water was dark blue, and quickly identified the beast as their own phone and electric line, probably slipping down through the clamps on the main cable.

Beebe told Hollister to continue the descent. They left the searchlight on and as the Bathysphere moved through 1,050 feet they marveled at the revelation of a dark world never before seen. Another squadron of hatchetfish moved into the narrow beam. They had been blue and purple lights in the darkness but in the light became tinsel-silver baubles bouncing in a synchronous tempo to an unheard beat. A mist of pteropods frisked in and out among the hatchetfish like gnats around flames.

Beebe asked Barton to shut off the light. A world of inky blueness returned in which constellations as bright as stars in a prairie sky formed and disappeared constantly. The Bathysphere was cooling rapidly, down to 72 degrees according to the thermometer, and the only sounds Beebe and Barton heard were the hiss of the oxygen and Hollister's voice piping faintly from the headphones as if from a hundred miles away. Neither the frantic tension of their first dive nor the emergency ascent of their second had prepared the men for the stark sense of isolation and danger that accompanied them as they neared the extreme reaches of sunlight.

At 1,100 feet, the sphere jerked to a halt while once again a hose clamp was tied on above, and Beebe and Barton took a moment to inspect the capsule that was keeping them alive. The door wasn't leaking, the oxygen tank was hissing, and the palm leaf fan was maintaining enough circulation to keep the air sweet. The walls of the sphere were dripping with moisture from the heat of their bodies condensing on the cooling steel beyond the ability of the calcium chloride to absorb it. Beebe suggested what he called a "grand shifting of legs" and the men enjoyed the brief pleasure of motion after being frozen in place for more than half an hour. Then they settled down, glanced at each other, and nodded at the same time. "Miss Hollister," Barton spoke up the line. "The Director and I will continue the descent."

To reduce condensation on the windows, Beebe wrapped his bandanna around his face like a cowboy in a dust storm. Then he crouched forward as the sphere began moving again. The searchlight was out, and he watched green, violet, and blue lights flash in no apparent pattern and then dissolve into the darkness. Beebe had seen bioluminescence in a few

jungle insects, in a tree fungus called foxfire, and in his deep-ocean specimens in the darkened lab aboard the *Arcturus*. But he was beside himself
with excitement as he witnessed its virtuoso performance in the darkness
of the deep sea.

Fluorescence or phosphorescence is produced by plants and animals
that absorb photons of light and then release them in darkness, such as
the algae that glow when disturbed in the wake of a ship. The lights in
the skin and organs of the animals that Beebe saw in the abyss were
bioluminescent—produced by chemiluminescence, a reaction inside
their bodies triggered by the interaction of two chemicals, one of which
contains oxygen. The most common reaction, and the one that was
turning hatchetfish into living lightbulbs in the Bathysphere window,
takes place when a chemical called luciferase triggers another chemical
called luciferin, which releases energy in the form of light. Both components in the reaction that produces bioluminescence come in several
different forms, which accounts for colors that span the spectrum from
the red to the violet extremes.

Beebe switched on the lamp again and began tracking random flashes
with the searchlight. He thought that he would always see a creature's
lights before it became a fish or a jelly in the beam. But at 1,200 feet, a
long, slender, golden-tailed serpent dragon, *Idiacanthus,* suddenly materialized in the searchlight without any previous hint of illumination as
though from another dimension. As he watched it writhe and turn in the
glare as though excited by the light, Beebe realized that the dragon,
which from his net samples he knew had at least three hundred light-
producing organs, could either illuminate itself at will or was triggered
by some unknown stimuli. It swam out into the darkness, still without a
trace of its beautiful fireworks. Beebe described everything he saw to
Hollister leaning into Barton's microphone, and the frightening silence
of the sphere was banished by the enthusiasm in his voice.

Then, between 1,250 feet and 1,300 feet, Beebe saw no shrimp, no
hatchetfish, no sea dragons, no larger creatures at all. Beebe had looked
at seawater with a microscope, so he knew that it was never truly empty.
Every square centimeter contained thousands of microscopic organisms,
some of them plants called phytoplankton, some of them minute animals called zooplankton. Thousands of species of diatoms, obelia, bristle
worms, copepods, radiolara, and the tiny larvae of jellies and crustaceans
were visible only as motes in the beam of the searchlight. But it was as

though a tunnel had been cut through the sea into which no life other than the plankton was permitted. Beebe pointed it out to Barton, they flashed the searchlight, saw nothing but its beam piercing the darkness, and together they came up with the theory that some kind of thermocline or temperature barrier was responsible. As suddenly as it had begun, the void gave way to life again, only this time it was fireflylike sparks that completely disappeared when they turned the light on. Whatever was making the sparks was too small to be seen in the beam. Each time the Bathysphere bounced to a stop while the Okonite hose was clamped to the lifting cable, Beebe and Barton were startled from a trance induced by the beautiful world outside their hissing, ticking little globe of steel.

Finally, at just over 1,400 feet and forty-four minutes into the descent, Beebe told Hollister that they had gone deep enough for that day and asked her to prepare the crew above for the ascent. He had no premonitions, but they had been crouching in the Bathysphere for more than an hour and fatigue was beginning to set in. They would go deeper the next time, and a quarter mile had a nice ring to it. As they hung at exactly 1,426 feet, Beebe was riveted to the glass watching a tiny, semitransparent jellyfish throb slowly through the beam of the searchlight. He gazed straight down into the darkness falling away from the bottom of the sphere and told Hollister that it looked like the blackest black of hell.

The naturalist was jarred from his reveries by Barton's voice saying that he had calculated that the pressure on the three-inch-thick window against which Beebe's face was pressed was more than 650 pounds per square inch, and that over 6,500,000 pounds rested on the entire surface of the sphere. Under normal circumstances, those numbers would be as meaningless as saying that the Andromeda nebula was 900,000 light-years away from Earth, but every digit was starkly real to Beebe at that moment. The creatures outside the Bathysphere didn't feel the pressure because they were filled with water that had already been compressed, but the men inside, carrying their water and gas from a far less pressurized world, would not last a microsecond in the abyss. Beebe knew that there was no chance that they would drown if the windows or the steel ruptured, since the first molecules of water entering the sphere would fly through their bodies like high-powered rifle bullets.

As they dangled at a depth of 1,426 feet for thirty seconds, Beebe was astonished by the circumstances in which he found himself. A transparent bit of sand that had been melted and re-formed into a piece of glass was holding back nine tons of water from his face, and he was swept with a wave of emotion and a real appreciation for where he was. The barge slowly rolled on the surface a quarter mile above, the long cobweb of a cable led down to a lonely sphere in which two human beings sat as isolated as a lost planet in outer space. He was overwhelmed by the privilege of being able to peer out and actually see creatures that had evolved in the blue-blackness of a midnight which, since the ocean was born, had known no daylight. The words of Herbert Spencer formed in Beebe's mind: he felt like "an infinitesimal atom floating in illimitable space." With that phrase, he also wrote in his notebook: "Am writing at a depth of a quarter of a mile. A luminous fish is outside the window." And he signed his name with a flourish.

Then Beebe and Barton felt the weighty sensation of ascent, and they rose through layers of water that were beginning to seem oddly familiar, continuing to note the creatures and conditions through which they passed. They made the return trip in forty-three minutes, stopping briefly every hundred feet while hose clamps were removed from the lifting cable. The strange, empty zone between 1,300 and 1,250 feet was still there. Everywhere else, jellies, squid, hatchetfish, sea dragons, and unidentifiable luminescence ghosted past the little windows, seen by humankind for the first time though they had shared the planet with us for hundreds of thousands of years. Beebe and Barton instinctively relaxed during the ascent that brought them steadily out of darkness and into their own world of sunlight.

At one thousand feet, Hollister sent down word that she could see a bird flying in a tight circle over the barge and a moment later identified it as a herring gull. Beebe replied that he had made a note of the gull and told Hollister that he was now qualified as the first ornithologist who had ever made a submarine bird note. Hollister laughed. Seconds later, he told Hollister to correct the record because he remembered that he had seen a pair of penguins in the water while helmet diving off the Galápagos and had made a note on his zinc underwater tablet. Even Barton laughed.

They were on the surface with sea foam washing over the Bathysphere at 11:30; twenty minutes later, Barton and then Beebe emerged from the Bathysphere after two hours and two minutes inside. Both were wobbly-legged and had brutal headaches from the oxygen, but they were exhilarated because they had survived another descent and proven beyond a doubt that the Bathysphere was a precious tool for scientific discovery. Beating records or "becoming the-first-white-man-who-had-ever, etc.," Beebe insisted, were incidental. Fish he had seen only dead or damaged at the surface were now alive in the mind of humanity, and from now on, when he looked at the deep-sea treasures in his nets he would feel as an astronomer might who returns to his telescope after having rocketed to Mars and back, or like a paleontologist who had seen his fossils alive.

While the excitement of the quarter-mile descent still crackled across the *Ready*'s deck, Beebe called for quiet and announced that he and Barton had a surprise for Gloria Hollister. She was thirty years old that day, and in honor of her birthday and to thank her and Tee-Van for their service to the department, they would make the next dive in the Bathysphere. Hollister had been so caught up in the excitement and tension of her role as the life link to the Bathysphere that she had completely forgotten the significance of the date. The sea conditions were still perfect, the sphere was in top shape with plenty of oxygen for a short dive, and Hollister and Tee-Van didn't have to be asked twice. They dove into the sphere. Barton and a crewman swung the hatch cover into place and pounded home the main bolts, Beebe said farewell through the center hole and twisted in the wing bolt.

Inside, Tee-Van adjusted the oxygen flow and clamped the earphones onto his head, while Hollister took Beebe's usual position at the center observation window. She was an ichthyologist, and though she had made dozens of helmet dives to depths of sixty feet, this was a dream come true. She felt much safer and calmer than she had on the *Ready*'s deck listening to the disembodied voices of Beebe and Barton, and knew that her entire life had led her to precisely that moment.

For the next half hour, Hollister dutifully called up her observations of pteropods, shrimp, jellies, and fish to Beebe on the topside telephone, and she and Tee-Van joined the exclusive club of deep-ocean explorers. Hollister marveled at a white, tissuelike creature she knew to be an eel larvae called leptocephalus which was infinitely more graceful in its own world that it had been in her laboratory aquarium. Their dive ended at 410 feet, deep enough for Hollister to set the world depth record for women but not too deep to worry Beebe that his magnanimous gift might become a tragedy. Hollister and Tee-Van pleaded for another hundred feet, but Beebe wouldn't budge. He realized as he stood dry on the deck that a descent in the Bathysphere was more dangerous but nowhere near as frightening as staying behind with someone you loved in the depths below.

The next day, Beebe filed his own story for the *New York Times,* which ran on the front page.

BEEBE AND AIDE DESCEND 1,426 FEET INTO SEA
IN STEEL SPHERE, PHONING OBSERVATIONS TO TUG

BERMUDA, JUNE 12.   I descended with Otis Barton yesterday in a steel ball five miles south of Nonsuch Island to a depth of 1,426 feet, checked and double-checked.

We were lowered at fair speed from the deck of the tug. Barton attending to the telephone and being clearly heard by Gloria Hollister, who was taking notes on deck. I was watching the observation window. The cable meter registered 1,426 feet and the measured lengths of cable showed 1,428 feet where the pressure was 652 pounds to the square inch and the weight of the water on the sphere 3,100 tons.

We flew the flags of the Explorers Club and the New York Zoological Society.

A dead fish fastened to the sphere attracted the deep-sea fish in the vicinity, many luminous ones and shrimps were readily observable, attesting to the scientific value of the apparatus. All the light rays except the blue and violet at the end of the spectrum were cut off and it seemed brilliant outside, but the light was insufficiently strong to read the oxygen gauges within.

The only discomfort we felt was the reduced pressure when, after one and a half hours from the time we had left the deck, the sphere was opened and the pressure relaxed. It had increased due to the oxygen released from the tanks, only the carbon dioxide having been absorbed by the sodium. Otherwise we were both perfectly all right.

John Tee Van and Gloria Hollister descended 400 feet and observed fish formerly known only at greater depths.

Dozens of other accounts appeared in newspapers from Hamilton to San Francisco, and most of them added Beebe's intention to dive to a half mile before the summer was out.

But the diving season of 1930 lasted only nine more days. They sent a movie camera down in the empty sphere to 2,450 feet and triggered it with a timer. The film was blank, but the sphere returned dry. Beebe and Barton made two more descents to eight hundred feet, both of which ended because of static on the telephone line. Beebe also wanted to demonstrate that the Bathysphere could be used for exploring shallower water, so they made four descents along the insular shelf to 120, 80, 350, and 210 feet, which Beebe called contour dives. On two perfectly calm days, Beebe brought the *Gladisfen* and the *Ready* as close to coast as he dared, lowered the sphere into the water, and allowed the barge to drift away from the shore, towing the submerged Bathysphere with it. On the first two contour dives, the flow of moving water spun the sphere like a yo-yo on a string, but Barton came up with a wooden rudder that kept the windows facing forward.

Then, on June 21, a slow-moving storm drove them off the ocean for a week. Barton was on top of the world as he left St. George on the steamer back to New York and promised to return for more diving in August. Meanwhile Beebe had the Bathysphere unloaded from the *Ready* and stored in a tin hut at the Darrell and Meyer wharf. He then got busy entertaining patrons and distinguished visitors, including William Gregory from the American Museum of Natural History and Gregory's protégé, Jocelyn Crane, a young biologist and the daughter of one of the society's patrons. Beebe took Gregory and Crane helmet diving off Gurnet Rock, and the pair settled right into the routine on Nonsuch. Gregory stayed a month, and Jocelyn Crane became a member of Beebe's staff. She would never leave him.

In mid-July, W. Reid Blair, the tightly wound director of the Zoological Society and nominally Beebe's boss, came to Bermuda on an inspection tour. Beebe operated independently and drew his power in the society from his fund-raising prowess and his reputation as an author and naturalist, but his popularity rankled a few members of the administration, including Blair. When he arrived, Blair sent word to Beebe that he would not be accepting his invitation to visit Nonsuch for a tour and some helmet diving and asked Beebe to meet him in Hamilton to discuss society business. Beebe complied with his usual deference to authority, but came away from his meetings with Blair with the distinct impression that the man thought Bathysphere descents were stunts that would feed Beebe's popularity but contribute little to the scientific understanding of the ocean and its creatures. When Blair left, the Bathysphere was in storage in St. George and Beebe wasn't at all sure it would ever again carry him into the abyss.

As though Beebe had used up all his good luck in his record-setting descents, a plague of misfortune settled over Nonsuch Island and its band of adventurers for the rest of the expedition. The weather couldn't have been worse, with storm following storm and one of them developing into a hurricane that kept everyone on the island awake for two days feverishly working to save the buildings. The *Skink* developed a bad leak that took two weeks to find and repair, and the drum of the *Arcturus* winch cracked under the lateral strain of the cable while Beebe was towing nets two miles down in the cylinder, killing all hope for more Bathysphere diving that summer. Illness and injury also paid calls on Nonsuch. Jack Connery fell between the *Skink* and the wharf and broke a bone in his back. Tee-Van developed an obstinate cough and ran a temperature for a week, so Beebe sent him back to New York. One of the field assistants, eighteen-year-old Patten Jackson, died of appendicitis, and Gloria Hollister collapsed from exhaustion and had to leave for a rest in the village of Somerset to the west of Hamilton.

Other than the exhilaration of the two weeks in June, when the Bathysphere descents allowed Beebe, Barton, and the Crowd on Nonsuch to breathe the rare atmosphere of heroic success, Beebe's birthday party was the highlight of the expedition year. Since his Guiana expeditions, one of Beebe's favorite ploys when morale was low was to announce that it was his birthday—of course everyone knew it wasn't—and demand that everyone be in costume as an animal or a dancer or a Bohemian by dinnertime. July 29, though, really was Beebe's fifty-third birthday. Blair had left the week before trailing his wake of skepticism, the Bathysphere was gathering dust in storage, and he had no idea if he would be able to find the money to resurrect it. But when Beebe sat down at his desk on Nonsuch to write about the party in his diary before he turned in that night, he still felt like a lucky man.

A most wonderful birthday party. After cocktails, we had a good dinner. Everyone was dressed up as a pirate, with rags and knives and bloody sabers. They gave me a cedar box with a few gifts. Then two boys got up, came behind me and blindfolded me, and dragged me down the path and onto the Skink. *Everyone got on and we pushed off. It seemed to me as if I was going to St. George but we went on and at last I was pushed out along a plank and got ashore at the inside of Castle island. I was led ashore and suddenly the St. George Orchestra struck up and we marched up to where the moat of the castle was. There Greg made a long address and the boys gave a wonderful performance of diving with the helmet and everything was faked and the ladder led into the dry moat. Here a boy tied a weird fish on to a line and I pulled them along and at last a boy with a 13 foot shark came up. Then Perkins pulled up a bottle with a note in it which was tossed to me. It was an old map of the island with an X-marked treasure. I went on and followed the line, and in a bed of cactus found a big box and we carried it down to the dungeon which was lighted with candles and lined with fish net, and here we ate cake and beer and wine and I opened the box and found many gifts and the orchestra played. We felt that it equaled last year's party.*

INTERLUDE, 1931

I don't like belonging to another person's dream.

Lewis Carroll, *Through the Looking-Glass*

The two men who had descended a quarter of a mile into the Atlantic and gazed through ultramarine twilight at living phantasms returned to New York on a wave of publicity that surprised both of them. Newspaper stories and newsreels always mentioned their discovery of creatures and the weird light in the abyss, but the meat was in the danger. "A deck winch sputters. A long-armed derrick swings the bell over the rail. The cable is paid out," wrote a typically sensationalizing reporter in the *Sunday World Magazine*. "Like a corpse, the sphere and its load of living men vanish. Accompanied by the funereal drone of the winch, the cable rolls off the drum. The expedition is off! . . . Two hearts leap inside the bell. They sink beneath all known depths. Eyes afire, Beebe pressed against the quartz glass port and beholds a swimming pageant."

Diving in the Bathysphere was an act of courage or insanity that ranked Beebe and Barton among the outrageous daredevils who were capturing the imaginations of a nation laboring under the crushing burdens of the Depression. In hard times, people with nothing to lose but their lives lined up to attempt to drive, fly, and sail faster and farther in pursuit of new records, delivering daily rations of élan, recklessness, and often catastrophe to the newspapers and newsreels. During the very week that Beebe and Barton made their first descents, Major Sir Henry O'Neal Seagrave, who had already hit 231 miles per hour in a car on the sands of Daytona Beach, was killed trying to set a new water speed record when his boat, the *Miss England II*, flipped over at a hundred. Seagrave survived insensate for a few minutes with broken arms, ribs, and legs, and regained consciousness in the arms of his wife moments before he died. His last words were "Have I broken the record?"

The sky, too, was alive with firsts, highests, and fastests. Hugh Herndon and Clyde Pangborn flew nonstop from Saishiro, Japan, to Wenatchee, Washington, in forty-one hours and thirteen minutes. Wiley Post and Harold Gatty flew all the way around the world, landing for fuel and food, in 8 days, 15 hours, 51 minutes. U.S. Army Captain Hawthorne Gray set an altitude record in an open-basket hydrogen balloon in 1927, but was denied the official mark because he had to bail out with a parachute on the way down. On his next attempt, Gray made it to 43,000 feet before plummeting to his death when the basket thudded into a pasture near Sparta, Tennessee. He had made his last logbook entry at 40,000 feet in the barely legible scribble of an oxygen-starved man, but his recording altimeter showed that he had reached 42,740 feet before the balloon began its descent. Then Apollo Soucek, breathing from an oxygen tank and wearing electrically heated goggles, coaxed his airplane to 42,500 feet and back, breaking the altitude record.

Hawthorne Gray's sacrifice proved that a human entering in the hostile environment of the freezing, anoxic stratosphere was going to need more protection than a warm flight suit, goggles, and an open basket. Without a sealed, pressurized cabin, lungs cease to function somewhere between 40,000 and 50,000 feet, but nobody knew the upper limit until Gray used his own body to test that extreme. On May 27, 1931, Auguste Piccard and Paul Kipfer stood on Gray's heroic shoulders when they squeezed into an aluminum sphere eighty-two inches in diameter and lifted off from a field near Augsburg, Germany, under a towering black bag of hydrogen. The capsule in which they rode bore a striking resemblance to the Bathysphere; it weighed only three hundred pounds, but its thin aluminum skin was strong enough to contain a breathable atmosphere of oxygen scrubbed clean by soda lime that could keep two people alive above 40,000 feet for ten hours. They reached an altitude of 51,783 feet and survived.

Americans found vicarious relief from economic disaster in the exploits of speed demons, high-wire walkers, marathon dancers, and magicians who escaped from mortal bondage. A couple of men sealed inside a steel ball on the end of a cable descending into the cold, dark, deadly depths of the ocean fit right in. Gray, Piccard, Kipfer, Beebe, and Barton were ordinary human beings moved by desire and circumstance to explore the unknown. Every frontier is freighted with danger, but it is hard to say exactly where conscious courage enters into the decision-making process to move into new territory. Or even if it does. Many men and women who run into burning houses, dive into the abyss, or ride into the sky in wicker baskets will tell you they don't consider themselves to be particularly or extraordinarily brave and were not thinking of themselves at the moment of mortal truth. There have been some systematic studies of courageous acts, but we remain so baffled and awed by selfless impulse that we pack it into cocoons of mystery and allusion best served by poetry, novels, legends, and myths.

In the fourth decade of the twentieth century, when the lives of William Beebe and Otis Barton intersected with the technology to build the Bathysphere, the mechanisms for the anointing of heroes on a mass scale had blossomed as never before. Newspapers, books, and magazines remained fundamental to creating them, but Thomas Edison's wondrous motion pictures and David Sarnoff's radio network added a new degree of magnification to the exploits of Piccard, Kipfer, Beebe, Barton, and other death-defying explorers. The new movies of 1931 and the spring of 1932 included Fritz Lang's *M,* Charlie Chaplin's *City Lights,* the Marx Brothers in *Monkey Business,* Bela Lugosi in *Dracula,* and James Cagney in *The Public Enemy.* The surprise cinematic hit of the season, though, was a true-life adventure movie, *Bring 'Em Back Alive,* which drew 82,660 people during its first two weeks at the Mayfair Theater on Broadway, eclipsing the record set a few months earlier by Boris Karloff in *Frankenstein.*

Bring 'Em Back Alive was based on the best-selling book by Frank Buck, an animal collector from Texas. The movie depicted battles between tigers and leopards, pythons and boars, and other fights that crackled to life on the big screen, giving audiences a tantalizing taste of danger with no mortal consequences. As the Depression deepened, Otis Barton was faced with finding a way to make a living. He saw Frank Buck's film shortly after it opened and said to himself, That is the way to make money as an adventurer. He wasn't down to pocket change, but his dividend checks had stopped coming, he was eating into his capital, and whatever he spent from then on had to carry him into a profitable independent career of some kind. In the summer of 1931, the city suffered through a heat wave made all the more miserable because tens of thousands of men there had lost their jobs since the crash, despite Herbert Hoover's assurances that hard times were over and the nation was back at work. Around the rest of the country, another 10 million were unemployed and the Depression was getting worse. Small fortunes, like Barton's, and scientific expeditions, like Beebe's, were on borrowed time. Since Barton had stowed the Bathysphere in the warehouse on the Darrell and Meyer wharf and returned to New York in the summer of 1930, underwater moviemaking had become his next big idea.

The first thing he did to plot his new course was to give the Bathysphere to Beebe and the New York Zoological Society, shedding responsibility for the major costs of maintaining and operating the sphere with the condition that he be part of any future deep-diving expedition. Barton began to think of the Bathysphere chiefly as a camera blind from which to shoot movies of deep-sea creatures that would be the aquatic equivalents of Frank Buck's sensational forest and jungle fights. He bought several of George Eastman's new cameras, made two trips to the Caribbean to test them in underwater housings while helmet diving, and spent some more of his money buying his way into the film community by hiring big-name cameramen and directors for his tests. Barton also went to work designing a system to make the Bathysphere a better camera platform. He knew he needed a brighter searchlight, more power to run it, and a third window in the sphere through which to point his camera. These details occupied him through the spring of 1931 when he wasn't vacationing or testing his cameras off Nassau.

Barton was also anxious to shake off what had increasingly become his role as a second banana to Beebe. In most of the publicity and celebrations of their achievements of June 1930, Beebe was cast as the intrepid explorer and Barton as the sidekick. Though Beebe always acknowledged Barton's role in designing and diving in the Bathysphere, he simply didn't care much for Barton and instinctively distanced himself from his young collaborator. Barton could sense all of this and knew that without his money and his willingness to spend it building the sphere he would not have had a prayer of becoming a member of Beebe's Crowd.

While Barton was working on his metamorphosis into an adventure filmmaker and getting out from under the shadow of his former idol and somewhat estranged partner, Beebe remained Beebe. Barton had given him the Bathysphere, and Beebe was convinced that the news value of deep diving would more than make up for its cost in the kind of publicity and patronage that opening the abyss to science would attract. Early in 1931, one of Beebe's most loyal supporters, Mortimer Schiff, died and left the Zoological Society $50,000, part of which assured the expedition's return to Bermuda that year. Beebe also raised cash for salaries and ongoing expenses for his Department of Tropical Research from Bill Boeing, Perkins Bass, and Frederick Sturges, Jr. John Roebling gave him three thousand feet of quarter-inch trawling wire for his nets, George Eastman gave him $100 worth of film, and Gristedes of Manhattan kicked in a shipment of groceries.

Beebe and his staff, minus Barton, who was still listed as Associate in Charge of Deep Sea Diving, went back to Bermuda on May 6, 1931. That summer, the Beebe expedition hit snag after snag but still collected ten thousand specimens in their nets and traps, including the season's prize, a sacopharynx eel fifty-five inches long with a scarlet light organ near its tail. Beebe, Tee-Van, Hollister, and the new addition to the inner circle, Jocelyn Crane, spent a lot of time helmet diving and compiling a catalog of 320 species of near-shore fish that they planned to publish in a definitive book. In addition to their usual regimen of collecting, identifying, and preserving specimens and observing them in the near-shore shallows, Beebe's Crowd went to dinner and danced at the St. George Hotel and Government House, mounted the occasional costume party, and played volleyball and deck tennis.

The 1931 season, though, was burdened by more bad luck than good. In midsummer, just as the work on Nonsuch was gaining momentum, Elswyth called and told Beebe that his father had died. He took the next boat for New York, buried Charles Beebe in East Orange, New Jersey, and was back on Nonsuch a week later with "no Dad for the first time," he noted in his diary. (His mother, Henrietta, had died in 1925.) The day after Beebe returned, the *Gladisfen* sprang a leak. A week later, though, she was ready for sea, so Beebe went back to work towing nets through the cylinder, spending weather days in the lab, and keeping things flowing in his river of inquiry.

In June, Otis Barton was in the Caribbean testing his underwater cameras, but told Beebe he might be able to come to Bermuda in July or August for more diving. The Bathysphere and its rigging were in St. George, where Tee-Van inspected them and reported that they were in perfect shape. But on June 17, during a deep tow through the cylinder, the drum of the *Arcturus* winch, which was mounted on the *Gladisfen,* cracked again; a stronger, specially built replacement did not show up in Bermuda until the end of July. By then, the vagrant thermals of late summer were spawning weekly storms, four of which put Bermuda on hurricane alert, and the weather never again broke long enough to settle the sea into the calm surface necessary for launching the Bathysphere. Even when the wind did let up for the odd day or two, something else seemed to go wrong. Beebe's sinuses were killing him, and for two weeks in mid-July he was forced into light duty and bed rest. The *Skink,* which Beebe used for towing and trapping in shallow water, had to be completely overhauled.

Beebe's first major article on the Bathysphere descents of 1930 appeared in the June issue of *National Geographic.* He called it "Round Trip to Davy Jones' Locker," and the story, with all the other publicity attending his record-setting dives, had the desired effect on his treasury. Still, he spent a lot of time shoring up funding for the 1932 expedition and the expenses for another year for the Department of Tropical Research. The money-raising power of the Bathysphere was far outstripping Beebe's expectations, and he capitalized on the headlines and dozens of interviews by promising a descent to a half mile. His vow had a sensational ring to it, and a half mile was also near the maximum depth possible with a cable of 3,200 feet. Beebe made sure every reporter knew about his goal, because the few remaining men with any real money in New York might be interested in vicariously riding along with William Beebe as he set records, made headlines, and explored the new frontier of oceanography.

Beebe was right. In July and August, George F. Baker, Elihu Root, Jr., Frederick Sturges, and Henry Palmer made trips to Bermuda to visit the expedition. One after another, they anchored their yachts in the spectacular turquoise shallows of Castle Harbour, toured Nonsuch Island, and went helmet diving. The tycoons reciprocated by having Beebe and his Crowd aboard for drinks and dinner, and wrote checks on the spot for 1932. Back in New York, David Sarnoff at the National Broadcasting Company had expressed an interest in actually putting the exploits of the Bathysphere expedition on the radio. And Beebe's tried and true patrons, Mortimer Schiff's son John, who had taken over his father's philanthropic enterprise, Boeing, Bass, and Williams, were promising less money than in previous years, but they were still true believers. And on the day of his real fifty-fourth birthday, Beebe landed a whopper.

Colonel Edwin Chance, a Philadelphia banker, arrived on Nonsuch with his wife on July 29, and Beebe turned on the bright lights. They spent the day helmet diving off Gurnet Rock and in the evening settled into the big parlor of the main house to watch movies of the Bathysphere dives of 1930. Chance loved being around Beebe's playful energy, and before stumbling by lantern light down the dark trail to return to his yacht the *Antares,* he leaned over the dining room table and wrote Beebe a check for $500. The next day, after Beebe won his match in a tennis tournament that set him up for the championship the following week, he, Tee-Van, and "the girls"—Hollister and Crane—chugged over to the *Antares* for dinner. Chance was an effusive host aboard his perfectly appointed yacht, and he and Beebe spent the evening in fluid conversation, some of which centered on the banker's skepticism about the depth of hard times. Chance maintained that the Depression didn't really exist at all, that America was like a little boy who has been feeding on ice cream having to go back to oatmeal.

The next day, Chance's wife sent Beebe her analysis of his handwriting from a note he had written to accept the dinner invitation aboard the *Antares.* In Beebe's handwriting, she said, was evidence of "concentration, enthusiasm, zeal, physical activity, affection, sympathy, friendliness, straightforwardness, honesty, boldness, kindliness, scientific mind, sagacity, curiosity, spirituality, broad mindedness, sense of justice, clearness of ideas, judgement, intuition, self respect, bravery, activity of mind, health, literary taste, command of language and expressiveness." The Chances blended right in with the Crowd on Nonsuch, and on the night of July 31 they joined the party to celebrate Beebe's birthday two days late. Arthur Tucker lit South Beach with bonfires, a band from St. George played for dancing, and Tee-Van staged a burlesque on the discovery of Bermuda. Before he left, Edwin Chance promised Beebe that he would give him at least half of the money he needed to dive again in 1932.

Part Two

BEYOND SUNLIGHT

Nine

RENAISSANCE

It's a long way from amphioxus,
It's a long way to us.
It's a long way from amphioxus,
To the meanest human cuss.
Good-bye fins and gill slits,
Hello skin and hair.
It's a long, long way from amphioxus,
But we got here from there.

Sam Hinton vs. an anonymous paleontologist
(to the tune of "It's a Long Way to Tipperary")

On the morning of August 31, 1932, William Beebe, Jocelyn Crane, Gloria Hollister, and John Tee-Van scrambled up the sea-slick ladder from the *Skink*, which was bobbing like a cork in a churning washtub alongside the Darrell and Meyer wharf. They had come from the expedition's new headquarters at the Bermuda Biological Station, and their slow passage from Ferry Reach into the belly of St. George Harbour had drawn a knot of curious bystanders at the dock by the time they threw up their lines to clutches of men waiting to help. Since the summer of 1928, Beebe and his assistants had been coming and going in full view of the St. George waterfront, and over the past four years their glamorous adventures had become a source of local pride. Building diving helmets had caught on as a local craze among teenaged boys, and a buzz ran through the islands whenever a report about the Bathysphere made its way into the chatter and gossip that were so much a part of every languid tropical day. That day, word had circulated that the strange craft which had been hidden from view in a shack on the wharf for two years would be brought to life again.

Otis Barton, wearing a casual tan field shirt and olive drab trousers instead of his machine shop greens and looking much less like a kid than he had at the end of June 1930, was among the people milling around on the dock. He had entirely lost his hair save for a tonsure on the sides and back of his head almost identical to Beebe's. Barton had arrived two days before on the *Queen of Bermuda* with his crates of cameras, diving helmets, and, in a special case resembling a treasure chest, a third quartz window for the Bathysphere, all of which formed a substantial pile on the wharf. Beebe and Barton had seen very little of each other since parting in the summer of 1930 and they greeted each other as though at a diplomatic reception. Barton shook hands with Tee-Van, nodded at Crane and Hollister, and introduced everyone to his personal cameraman, Laurance Fifilik, who could not contain his excitement over the impending adventure and was chattering at anyone who would listen.

Their Victorian manners got them through the introductions, but Beebe and Barton were uneasy in each other's company. Barton was not happy to be back in the shadow of the man whose name always preceded his in the newspapers, and he bristled at Beebe's instinctive dominance. Barton felt more comfortable with Tee-Van, so he sidled up to him for social shelter, which was especially welcome since Hollister and Crane were their usual dazzling, feminine selves and didn't want to reveal his awe. Beebe noticed everything about Barton's behavior, as he would that of a bird he was observing from a blind. There was simply something about the young man that did not suit him, some collection of irritating traits in this son of wealth and nephew of the president of Harvard. Despite the simmering acrimony between the two men, neither could afford to pass up the chance to dive again, Beebe for the sake of the very existence of his oceanographic expeditions, Barton for his nascent career as the Frank Buck of the abyss.

An awkward silence hung for a few seconds, then Beebe clapped his hands and led the way to a ramshackle corrugated iron hut about the size of a three-horse barn. The dockyard crew removed the front panel of the hut, forcing everyone to move back a few steps before they ducked under the low ceiling and gathered around the Bathysphere. There was a smell of decay and the sound of insects scuttling away from the brightness. A wedge of light from the sun over St. George Ridge fell on the sphere, and Beebe was startled to see that it looked more like the scarred, dull shell of an ancient tortoise than the craft in which he and Barton had made history. It should glisten, Beebe thought, and convey a readiness for supreme effort; but the Bathysphere was streaked with oil, grease, and rust under an ochre patina of dust. Beebe took out his pocketknife, flicked open the blade, scratched through the dirty white paint and smiled when he saw the silver-bright steel. He slapped the side of the sphere, then Barton moved up and did the same thing, and they both pushed on it without result as though to assure themselves that it still weighed two and a half tons. They moved around to the windows, knelt, and examined them. The one on the right side seemed slightly smoky, but the center one, through which they had seen the astonishing creatures of darkness, was crystal clear.

A clatter from the wharf broke the reverie, and everyone moved away from the sphere to stand outside. Two men dragged a block and tackle into the shed and more men removed the tin roof, exposing the Bathysphere to the sky, where the boom of the heavy dock crane was swung into place. To cut costs, Beebe had chartered the single tug *Freedom* instead of both the *Gladisfen* and the *Ready,* but her captain, James Sylvester, had assured him that the heavy oak mast, boom, and rigging from the old barge and the *Arcturus* winch could be mounted on her. The new mother ship was seventy feet long, with her wheelhouse and crew's quarters set aft and a broad, open foredeck. The *Arcturus* winch would be set behind the mast, its cable fed forward to the bow, through a pulley, then back to another pulley at the foot of the mast, and through a third pulley on the boom. It was the same rig, minus a second winch on the bow, vetoed by Captain Sylvester in 1929, but now it would have a stronger mast and boom and the Bathysphere weighed two and a half instead of five tons. The winch would be powered by the steam from the *Freedom*'s main boiler. The tug was more than twenty years old and showing signs of age, but the experience of the first season of diving led Beebe and Barton to believe that she could handle the jobs of lifting and lowering the Bathysphere and staying afloat.

The *Freedom* wasn't the only new wrinkle in the 1932 expedition. Beebe was in Bermuda this time for the sole purpose of attempting a dive to a half mile that would be aired live from the Bathysphere by the National Broadcasting Company. Though Nonsuch Island was still his, he simply didn't have enough money to mount a full-scale expedition of six months of diving, towing nets, and cataloging specimens. Edwin Chance had limited his gift at the South Beach birthday party the year before to Bathysphere diving, and the rest of the river of contributions that had blessed Beebe's adventures had dried to a trickle in the financial drought that had gotten so much worse than anyone suspected it could. Now even Chance admitted that he was wrong about the Depression being just a downtick in the economy, though he was doing well enough to keep his promise to Beebe. General contributions to the Department of Tropical Research dropped from $7,450 in 1931 to $1,750 in 1932. Worse, despite pledges to the contrary, direct support for Beebe's deep-sea exploration work fell from $27,600 to $2,200, most of which came from Chance. Mortimer Schiff's bequest of $50,000 was paid as promised, but every penny went to the general operations of the Zoological Society.

It was the financial disaster that Beebe, Hollister, and the others who suffered through winter lecture circuits for years had worked so hard to avoid, now visited upon them by circumstances entirely out of their control. For Beebe, nature was a haven free of much of the strain of real-world politics and economics until the crash left him scratching for money along with almost everyone else in the world. The fifty-four-year-old explorer who still packed around a copy of *The American Boy's Handy Book,* though, prized resourcefulness above almost all of his other carefully tended character traits. He soldiered on. As his bank accounts went into free fall, Beebe made sure his friendship with Edwin Chance remained intact by accepting Chance's offer of his yacht the *Antares* for a cruise from June 22 to July 29, just before coming to Bermuda for the Bathysphere diving. Beebe dubbed it the Fifteenth Expedition of the Department, and he brought along diving helmets, dip nets, hand drogues, specimen jars, and Gloria Hollister. In five weeks, the *Antares* sailed from Philadelphia to Bermuda, then to Grenada, Union, Tobago Cays, St. Lucia, Martinique, Antigua, Barbuda, St. Kitts, and Saba. Beebe and Hollister dutifully made a list of 125 species of fish along the way, and the sea air and leisure finally cured the acute sinus inflammation, fever, flu, and head-aches that had plagued Beebe since the end of March. His illness, which eluded a definite diagnosis and came with a cascade of symptoms that had confined him to his apartment for almost two months, was another reason he could not undertake a full season at Nonsuch, though he kept that quiet.

Without enough money or time to open his field station on Nonsuch Island, Beebe had to ask for help from another friend, Edwin G. Conklin, who was the president of the newly launched Bermuda Biological Station. The station was one of a wave of research institutions that were founded after HMS *Challenger* threw the door wide open for marine biology and oceanography. At the same time that discoveries were pouring in from the sea, Darwin's theory was ripening and transforming most open-minded students of life on earth into evolutionary biologists. Beebe numbered himself among them. The greatest object of his curiosity was the origin of the spectacular displays of complexity and refinement that he saw in nature. He and Conklin had become friends because of their shared fascination with the creatures of the sea, which both men believed were leading actors in the drama of life's beginnings. Though the collections and observations from Beebe's deep-ocean research were valuable, his passion for deep diving grew because he knew there were clues to the puzzles of adaptation, selection, and survival in the abyss. If life on earth proceeded from a source point through hundreds of millions of years of change, taking billions of forms, he rightly assumed that genesis was surely in the sea and not some mythical or even metaphorical garden.

Edwin Conklin was older than Beebe by fifteen years, a professor of biology at Princeton University and the author of a landmark book on evolution, *Heredity and Environment,* published in 1915. Beebe knew the book, of course, but became fully aware of Conklin's reputation as an evolutionary biologist ten years after it was published, when Conklin was widely quoted as supporting the teaching of Darwin's theory in public schools during *Tennessee* v. *John Scopes,* the famous Monkey Trial. Scopes, a high school biology teacher, volunteered to be arrested for teaching evolution as a test case to challenge a state law that banned the teaching of evolution in Tennessee classrooms. Clarence Darrow argued for the defense and William Jennings Bryan assisted the state prosecutor. A jury of twelve men, including eleven regular churchgoers, was startled when, in his closing statement, Darrow urged them to deliver a guilty verdict. (Darrow was counting on a review by higher courts.) The jury convicted Scopes and fined him a token $100. A year later, the Tennessee Supreme Court reversed the conviction on a technicality. Testimony and publicity during the trial tended to favor the defense and set the United States on a course of scientific authenticity in education. In support of Scopes and the teaching of evolution in schools, Conklin signed a formal resolution drawn up by himself, paleontologist Henry Fairfield Osborn, and eugenicist Charles B. Davenport, which was published in *The Daily Science News Bulletin.*

Beebe and Conklin first crossed trails in New York; they cemented their friendship in Bermuda, where they dove and collected together. Beebe was particularly vigilant about netting tiny fish-like creatures called amphioxi, or lancelets, and shipping them to Conklin for use in his research. Conklin's enthusiasm for these little creatures infected Beebe, who counted himself among the pioneers not only of deep diving but the systematic study of evolution.

Amphioxus are among the most primitive relatives of all vertebrates and contain evidence of the processes of evolution that are important to the study of life. Fish were the first true vertebrates, transforming the earliest prechordate body plans half a billion years ago into the bone and cartilage they would eventually pass on to amphibians, reptiles, birds, and mammals. More than half of all vertebrates, extant or extinct, are fish, including five distinct classes: *Agnatha,* the now-extinct jawless fish; *Placodermi,* the now-extinct armored fish; *Acanthodii,* the now-extinct spiny fish; *Chondrichthyes,* the sharks and ratfish; and *Osteichthyes,* the bony fish. Members of all five classes were alive during the Devonian period, known as the Age of Fishes, 405 to 345 million years ago. Fish continued to evolve for 140 million years until the Pennsylvanian period, after which only the sharks and bony fish continued the voyage into the future, thus making evolutionary room to increase the varieties among these efficient, adaptable survivors. Amphioxus remained, however, a near-vertebrate and one of the senior examples of the chordate experiment. Sometime before jawless fish took the big step of becoming true vertebrates, amphioxus, lampreys, and hagfish veered off into a niche that has kept the direct descendants of this primitive life-form alive for almost 450 million years. Beebe and Conklin were thrilled to be pulling this evidence of the dynamic past from the shallows around Nonsuch Island.

By the 1930 season, Conklin had become a regular member of the Crowd, and in 1932 he returned the favor of Beebe's hospitality. Conklin was a talented administrator as well as a competent scientist, and he had been instrumental in the establishment of the Woods Hole Marine Biological Laboratory in Massachusetts, one of four pioneering permanent ocean research bases created just before the turn of the century. In 1872, the Naples Biological Station had sprouted in Italy; then Woods Hole in 1888; the Hopkins Marine Station in Monterey, California, in 1892; and in 1903 the Marine Biological Association, which became the Scripps Institute for Oceanography in La Jolla, California. In 1903, Edward Laurens Mark of Harvard and Charles Bristol of New York University also joined forces to launch the Bermuda Natural History Society and began bringing student biologists to the islands. The Hotel Frascati in the village of Flatts was the first home of the society, where Mark and Bristol conducted informal summer gatherings and began building an aquarium. Then, in 1926, the Rockefeller Foundation, which also had endowed Woods Hole, donated seed money to create a corporation of 134 scientists from Canada, Britain, and the United States to establish the Bermuda Biological Station for Research, which was granted a charter by the Bermuda parliament that same year entitling it to own property. In 1928, the station had moved to its permanent quarters on property near St. George donated by Arthur Hollis, with a large hotel building, several cottages, a dairy, and stables for the necessary horses.

Edwin Conklin, who had been named president of the corporation, had casually suggested that Beebe apply to become the station's director. Beebe had been flattered but had declined, citing his commitments to the Zoological Society. In January of 1932, a taciturn English zoologist named John F. G. Wheeler took the director's job. Beebe was, however, a member of the corporation, and that same month, he and Elswyth were Conklin's guests in one of the cottages on the station grounds. Together they hatched the idea that, in light of hard times, Beebe and his expedition would use the laboratories and accommodations at the station to continue what was obviously going to be his scaled-down research in Bermuda.

Wheeler had other ideas. He and Beebe met at Wheeler's installation as director of the station, and during their brief conversation Wheeler asked Beebe if he would send him some specimens of plankton from the Nonsuch collection. In his note with the shipment, Beebe told Wheeler how happy he was that he would be able to work at the Bermuda station with the Northwest Cottage as his laboratory. Wheeler's reply, typed at the top of a plain sheet of paper, arrived in New York in mid-February.

> Dear Dr. Beebe;—
>
> The arrangement made by you to take over the N.W. cottage on the Station grounds must be suspended for further consideration. Owing to the failure of a letter from Dr. Conklin to reach me I have had no opportunity of expressing my views to him or even of hearing what arrangement has been suggested.
>
> I have received the box of plankton samples but have not had time to look them over.
>
> > Yours sincerely yours, [*sic*]
> > J. F. G. Wheeler

Wheeler also wrote to Conklin.

> We shall in future have wider aims than going down a full half mile underwater and I cannot see that the presence of Dr. Beebe and his people is going to advance our plans in the slightest degree.

Wheeler's frosty treatment was devastating to Beebe, for whom acceptance and respect in the scientific community were as essential as breathing. Not only were his plans for the coming summer cast into disarray, but his work was being dismissed as insignificant to science. With a stiff upper lip, though, he dashed off a note to Conklin:

I do hope I have not been indiscreet. I sent Wheeler a lot of plankton at his request and very casually mentioned my delight at being able to work for a time at the Station in the N.W. cottage. I shall be dumb after this & try to be tactful, for all I want is freedom to carry on my work. What a grand old world this would be if we could all be blessed with humanness, kindness & a sense of humor. You have more than your share.

But it wasn't only Wheeler who wanted to ostracize Beebe from the Bermuda station. After the opening ceremonies, another corporation scientist, Olive Earle, commented to Conklin: "a friend of mine went to the meeting of zoologists at the Waldorf [in New York] and came away with the idea that Dr. Beebe owned the new station."

And Henry Bigelow, one of the visiting scientists in Bermuda and a member of the station board of trustees, wrote Conklin to express his consternation about Beebe's presence: "Rumors seem to be flying around to the effect that Beebe and his train of servitors are apt more or less to monopolize the Bermuda lab, etc., etc., and that Wheeler is apt to be unhappy."

Some pragmatic insiders, though, offered a defense for Beebe's presence at the station. F. Goodwin Gosling, a rum merchant and member of the Bermuda parliament who had been instrumental in securing his use of Nonsuch Island, wrote to Edward Mark to point out the fund-raising power Beebe might bring with him: "If Beebe's expedition is successful, perhaps it may lead to a substantial grant being made by the Rockefeller Trust for . . . the Biological Station." Mark, however, disagreed with Gosling in words that crushed Beebe when they filtered back to him: "To tie up with such an explorer and exploiter would certainly kill our chances, if I can see straight."

Conklin left out the specifics when he reported the internal debate to Beebe, including the use of a childish sobriquet, "the camel," by several of his critics who thought it was a clever reference to Beebe's physical stature and the Arab proverb which cautions against allowing the "nose of a camel under the tent." But he could not sugarcoat all the disrespect directed at his friend. Beebe's innate arrogance, his immense popularity, and his lack of real academic credentials in marine science prompted Wheeler, Earle, Bigelow, Mark, and the others to exclude him.

At the end of February 1932, crestfallen by developments in Bermuda, Beebe had accepted an invitation to give a lecture in San Francisco and also spent a few days poking around in the tide pools on the coast to the south. He returned to Manhattan in late March carrying a sentimental but by then aromatic satchel of shells and starfish for Elswyth, and for their homecoming ceremony they sat on the floor of their apartment and sifted through the treasures, marveling together at the remarkable forms employed by mollusks and echinoderms to cope with the demands of making a living in the ocean. Beebe's childlike enthusiasm again charmed his young wife, who had first seen the ocean off California as a seven-year-old girl. It was the sort of thing, she thought, that made her and so many others adore her husband so much.

Then she told him the news from Bermuda, which was mixed at best and contained no adoration. The week before, Conklin had brokered a deal with Wheeler and the executive committee of the biological station that cleared the way for Beebe and his party to use housing and a laboratory there during the coming summer. Working behind the scenes, Conklin had mustered enough support among the committee members to prevent Wheeler from summarily excluding Beebe, but they did insist that their director should make all decisions about the specific assignment of quarters and labs. Wheeler subsequently denied Beebe the larger Northwest Cottage and installed him in the smaller house known as No. 31. Wheeler also prevailed in his demand that no mention of the Bermuda Biological Station be included with publicity about Beebe's work, especially his Bathysphere descents.

A few days later, Beebe collapsed with sinus pain, headaches, and fever, unable to manage even the most essential details of his life. Elswyth took over correspondence with Conklin to make the arrangements for the summer, and stayed with Conklin and his wife on a reconnaissance to Bermuda. There she did her best to mend fences with Wheeler, who was barely civil to her. But the Sixteenth Expedition of the Department of Tropical Research and the NBC broadcast were on for August and September.

After the dock crew lifted the Bathysphere from the storage shed to its temporary place on the deck of a big cargo schooner, the work of refitting, painting, and installing the new window was done while the mast, boom, and winch were being installed on the *Freedom*. Beebe and Barton went right to work scrubbing the outside of the sphere themselves, but they knocked off at midafternoon of that first day and, with the others, returned in the *Skink* to the biological station to watch a total eclipse of the sun. Beebe knew, of course, that the cosmos was utterly indifferent to him, but he was always attuned to omens in nature, and the last total eclipse had been in January 1925, the year and month Beebe left ornithology to explore the ocean. Clearly, this one was cause for celebration. He and his Crowd gathered in the yard of their lab and living quarters, and Jocelyn Crane and Helen Tee-Van served cocktails.

Beebe joked that the midday drinking was perfectly appropriate because the sun was about to go down, and there was no denying the festive atmosphere kindled by the resurrection of the Bathysphere. The gaiety of the afternoon was broken only briefly when Wheeler and his wife strolled past the cottage with their dog, nodding coolly but not stopping to join the party. As the sun darkened over Bermuda, a cool, vagrant wind arose and the water of St. George Harbour glistened in a weird, uncanny light that brought a shiver to Beebe, who was suddenly struck by "the loneliness of the earth in infinite space." He was an unrelieved romantic who found it easy to involve himself in the most elemental tragedies and miracles in nature, which is why he was so drawn to the calling of a naturalist.

The eclipse proceeded at the creeping pace of orbital mechanics into the real evening, when the party moved inside. Animated conversation about the next day's chores on the wharf eventually faded into sleep for everyone but Beebe. He was famous for his ability to nod off instantly, but on that night his thoughts wandered from Wheeler's condescension, to the eclipse, to the descent into the Atlantic he and Barton were about to make. Finally, the racket of an enormous flock of migrating plovers and sandpipers passing over the station shifted Beebe's thoughts to the wonder of such tiny creatures flying across seven hundred miles of stormy seas to reach Bermuda, and he slept.

The first thing in the morning, Beebe and Tee-Van went to St. George in the *Skink* under a light drizzle. There they met Barton, who as usual was staying at the hotel, and went to work installing the third quartz window. Barton had paid for the new glass and was the acknowledged engineer of the expedition, so he took charge of the job with Tee-Van as his assistant. Beebe watched every move, though, remembering the moment two years before at 1,425 feet when he pressed his face against the window with each square inch of it holding out six hundred pounds of water. Beebe would have been quite happy just to leave the steel plug in the third window housing, but Barton insisted that his dream of filming the creatures of the abyss could not be realized without the third porthole.

Barton fitted a wrench on the first of the ten bolts securing the collar to the window housing and loosened it with a hammer blow that echoed along the waterfront like a pistol shot. Down the dock, on the *Freedom,* Captain Sylvester and his men looked up from their work, then went back to their turnbuckles and rigging while nine more times the reports sounded. To those of them who had been at sea with the Bathysphere two summers before, the sound reminded them of sealing the main hatch and added a measure of excitement to their labor. The whole crew recognized its bravado and were thrilled to be part of it.

Fortunately, none of the window bolts was frozen, so in an hour the steel plug was out, the housing sanded clean, and a new layer of white lead laid down to provide a seal between the collar and the steel of the sphere. Barton and Tee-Van brought the case containing the new window aboard, pried off the lid, and lifted out the jewellike piece of quartz. It glistened with perfection, a marvel of clarity in a perfect circle with the heft of a large hunk of rock. Barton held it up and Beebe looked through it to see the white roofs and pastel awnings of the village of St. George across the harbor with absolutely no distortion. Gently, Barton eased the window into the channel in the housing, wiped off the white lead that oozed around its edges, and reseated the steel collar on the ten threaded studs. Tee-Van tightened the bolts and seated them with careful taps of his hammer. The three men crouched together to inspect the window for cracks and found none, and for the first time the Bathysphere had the three eyes of its original design.

Ten

DISASTER

*The howl of the wind, the roar of the waves, the electricity and ozone
in the air excited us beyond normal; no one thought of cocktails
at dinner time—Weather took their place.*

William Beebe, *Nonsuch: Land of Water*

The *Monarch of Bermuda,* with four men from the National Broad-casting Company among her passengers, was one of the last ships into St. George ahead of a slow-moving hurricane that was already knocking the tops off waves on the afternoon of September 4. Director George McElrath, announcer Ford Bond, and engineers W. C. Reisides and E. C. Wilbur were traveling with a ton of equipment, including radio transmitters, antennas, microphones, and cables. David Sarnoff, the founder and president of NBC, had orchestrated the deal for the undersea broadcast personally because he was a fan of Beebe's books and adventures. After closely following the sensational reports of the 1930 Bathysphere descents, Sarnoff saw an opportunity to use his revolution-ary radio network to capture the excitement of exploration as it was hap-pening and beam it into millions of homes. William Beebe was a perfect star in the emerging firmament of mass media, a celebrity whose name was already known to millions of nature lovers because of his best-selling books. Sarnoff's five-year-old network of stations radiated like an elec-tronic spiderweb from the Eastern Seaboard into the Deep South and west to cities on the Pacific coast. Since early August he had been telling his listeners that they could become part of the history being made in Bermuda simply by turning a dial in their parlors. William Beebe and Otis Barton, Sarnoff promised, would set a new world's record for deep-sea diving while the audience eavesdropped. The legendary naturalist and his aide would risk death to discover strange creatures never before seen by human beings. In return, the voices of Beebe and Barton would enter the homes of millions of people, and that kind of publicity would sell a lot of books and movie tickets.

In much the same way that Beebe and Barton were exhilarated by the challenges and rewards of the new science of deep-ocean exploration, the four radio men who shambled down the gangway at Town Wharf in St. George were awash in the possibilities of their own miraculous set of inventions. Their network broadcast from Bermuda would be the first ever from beyond the territorial limits of the United States.

Tee-Van met the *Monarch* and helped the radio men muscle their crates of gear aboard the *Skink* for the ride across the harbor to the *Freedom,* where the work on the mast, boom, and winch was almost finished. Ford Bond, a large, doughy man dressed in dapper knickers and knee socks, was particularly smitten by St. George and his imminent meeting with William Beebe, who had been one of his heroes since childhood. Bond was NBC's chief sports and special events announcer; after Sarnoff arranged the broadcast of history-making exploration, he and George McElrath took care of the details.

As the *Skink* puttered across the harbor, Tee-Van, ever the good host, delivered his usual recitation of St. George's history as the oldest permanent English settlement in the New World and pointed out the sites of the village dunking stool and a darkly celebrated witch burning. In minutes, the short trip was over and they nudged up behind the *Freedom* at the Darrell and Meyer wharf. Bond said something about rotting old hulks; Tee-Van assured him that in the tropics even young boats look bruised as he nosed the *Skink* to the dock ladder behind the tug.

When the radio gear had been safely stowed, Tee-Van took them to meet Beebe and Barton, who were busy painting the outside of the sphere a dark navy blue. The new color was Beebe's idea, better than white, he reasoned, for becoming invisible in the depths and less likely to spook curious fish. Though there was no denying the importance of the publicity from the radio descent, Beebe was fighting for his scientific credibility and had to deliver observations and discoveries that would impress not only the NBC audience but Wheeler, Blair, and the growing list of his detractors as well.

Tee-Van made the introductions, but Beebe didn't stop work for a conversation because the weather was obviously starting to close down ahead of the hurricane, still eight hundred miles southeast but moving steadily toward Bermuda. Beebe pointed out his shark-oil barometer hanging from the *Freedom*'s rail, and explained that local mariners had relied on them for centuries. The oil in a three-inch glass vial remained clear under normal conditions and clouded up when a storm approached. That day, the oil was milky white, an ominous warning of an imminent atmospheric disturbance that agreed with radio reports from Cape Hatteras and other stations to the southwest that were already in the grip of wild weather.

Bond and the engineers peered through the open hatch at the claustrophobic reality of the four-and-a-half-foot interior of the Bathysphere, and instantly felt profound respect for the courage of Beebe and Barton. Beebe broke the news that the hurricane, even a near miss, was going to stall the broadcast for four or five days because of rough seas, so the radio men might as well get comfortable and try to enjoy themselves. In two frantic days of work before the storm arrived, though, Beebe, Barton, and their crew managed to load coal for the boilers and get the Bathysphere ready to dive. Then they, too, waited.

Getting through a hurricane in Bermuda was not without its pleasures. There was the traditional rum punch, singing and storytelling by candlelight, and the groaning of the trees, the buildings, and the sea, which built to a frightening crescendo, a reminder to Beebe of humanity's fragile presence among the forces of nature. To a man who actively pursued the unknown and prided himself on survival under duress, violent weather brought an odd measure of comfort. While the radio men and Barton hunkered down at the St. George Hotel, Beebe and his Crowd manned the cottage on the bluff at the biological station and tended to roofs and debris for the better part of three days, secure in the knowledge that they had lashed down the Bathysphere amidships aboard the *Freedom,* battened the Okonite cable to the foredeck, and stowed everything else below. They had moved the tug into the lee of three ancient shipwrecks in the harbor and anchored her there in the heart of a spiderweb maze of ropes and cables. If the *Freedom* stayed afloat, which it had through dozens of storms and hurricanes, all would be well when the storm was over. Assuming, of course, that the radio men didn't just pack it in and catch the first ship back to New York.

Four days after a brisk southerly breeze announced the approach of the hurricane, its full force passed well to the south of Bermuda, and the wind and sea calmed on the evening of September 11. The *Freedom* and the Bathysphere had survived in the embrace of the old wrecks in the harbor, and the *Skink,* too, made it through the storm at anchor with a fleet of small craft in Mullett Bay, a tiny pocket of water protected by hills on all sides. There was a bit of bad news, though. When Beebe, Tee-Van, and Ford Bond went out with Captain Sylvester and his crew to move the *Freedom* from her storm mooring, they saw that the tug was well down on her lines, the result of the weight of the Bathysphere, the winch, the steel-and-Okonite cables, and an extra supply of coal for the diving operation. On the trip back to the wharf, they also found that she had shipped a lot of water that was sloshing in the unbaffled bilge, giving her an uncomfortable wallow.

Beebe did not like the idea of taking the *Freedom* and her precious cargo to sea in suspect condition, and he took a solitary walk on the dock to mull over whether to risk a dive. He watched the rest of the radio men arrive in a launch with Barton, Hollister, and Crane, saw the extra deckhands checking in with Captain Sylvester, and felt that a moment of truth had arrived. He had two hours of cruising out to deep water to determine whether conditions were safe enough for launching the Bathy-sphere, so he made the decision to put to sea.

Three miles from Castle Rock, the *Freedom*'s wallowing gait became a series of yaws and lurches as she encountered the leftover swell from the hurricane. Seasickness struck many of the crew, including Barton. Every-one braved the malaise with good cheer, bolstering one another with advice, teasing, and consolation as most of them visited the leeward rail. Just as Beebe began to realize that the sea was simply too rough to launch the Bathysphere, radio men or no radio men, Captain Sylvester sent a seaman down to the expedition party with word that the *Freedom* was leaking so badly that they were turning back. Water had already risen over the engine room floor and the pumps weren't keeping up with the leak, so there was no choice but to return to the harbor as quickly as pos-sible. The day was lost, and sinking was not out of the question, but the tug made it home. That afternoon, Sylvester sent down helmet divers to look for the hole in his boat: they searched for an hour before one of the men saw a good-sized gray snapper vanish through the hull right before his eyes. The gap was plugged, and the chief engineer spent part of the rest of his day unsuccessfully trying to hook the snapper with rod and reel while he supervised his men dumping water over the side with hand pumps.

On the morning of September 13, the *Freedom* nosed out of Castle Harbour and into the Atlantic again, but even without the weight of the water she was what a sailor calls a tender ride. This time, the pumps han-dled the water seeping through her sprung seams and unfound holes, and the sea had settled further in the wake of the hurricane, so spirits were high, especially among the radio men. Before leaving the dock, Beebe and Barton had stood alone next to the Bathysphere for some photographs, and while Fifilik clicked away, seemed to be arguing in hushed tones. After the pictures, Beebe announced that the plan for the day was to remove the chemical racks, oxygen tanks, telephone, and the lamp from the sphere and send it down to three thousand feet to test Barton's new window and the winch. If the sphere returned dry and intact, and the sea remained calm enough for a launch, they would attempt the half-mile descent the following day.

They had been cooling their heels for over a week already and the radio men had to conceal their disappointment, but McElrath marshaled them into enthusiasm for a dry run and told them to unpack their gear to rehearse for the real thing. The broadcast had been scheduled for September 11 and in New York, Sarnoff, an instinctive showman, milked the postponement because of a hurricane to build drama and tension. But it had to happen soon. Before the *Freedom* sailed, McElrath sent a telegram to Sarnoff advising him of the decision to spend at least that day testing the Bathysphere but assuring him that there was a good chance they would be on the air from the abyss soon.

After the tedium of two weeks of logistics, the claustrophobic uncertainties of the hurricane, and the ill-fated sojourn of the day before, the open Atlantic on the calm morning of September 13 was a tonic for everyone. The sky was overcast but clearing, a light swell remained to deliver an occasional mist of spray over the windward rail, and the aromas of the sargasso weed and the sea itself carried a measure of invigorating promise. The radio men were animated, glad the show was finally on the road; during the hour's ride to the deep water of Beebe's cylinder, they walked around the deck with Tee-Van, who showed them where they could safely set up their gear.

Beebe and Barton took the oxygen tanks from the Bathysphere, sealed the stuffing box with a short piece of Okonite because the test would be made without the power cable, closed the hatch, and hammered down the bolts. They were ready to launch when the now familiar angles of the lighthouses told Captain Sylvester he was six miles offshore and the deck shuddered as he pulled off the power and turned into the southwest wind. The *Freedom* protested the abrupt change of course with a typical groaning wallow to port, then settled and rode steadily enough so that the weight of the sphere and its cable swung out over the starboard bow would not capsize her. Beebe looked at Barton, turned his palms upward in a let's-do-it gesture, and gave a thumbs-up to Sylvester, who had come down to his station at the winch.

Sooty smoke belched from the *Freedom*'s single funnel as the stokers below shoveled coal in the boiler to get a head of steam up to drive the winch, and the test dive began. Since there was no scramble to attach the Okonite hose to the lifting cable, the descent for those on deck was just a matter of watching the meter gauge tick off the depth and making sure that the winch didn't foul or freewheel. After the dark blue Bathysphere faded to turquoise and then vanished completely about a hundred feet under the keel, the faces peering down from the *Freedom*'s rail saw nothing but the filament of cable slipping into the sea, heard the growl of the winch unwinding against the back tension that would keep it from spinning free, and felt the slap of the light swell on the side of the tug.

What they could not see, though, was that at about six hundred feet, when the pressure on the steel and glass reached about 250 pounds per square inch, a needle of water broke through the seal in the housing of Barton's new window and jetted across the sphere to the opposite wall with enough force to nick the metal. That minuscule needle of water also began eroding the white lead between the window collar and the glass itself, so that the flow had increased a hundredfold by the time it reached 1,200 feet. Almost certainly, the sphere was full of water at two thousand feet but no one topside could possibly know until, when the meter gauge read three thousand feet, Captain Sylvester's men dogged down the winch. At this point, the water inside the sphere weighed an additional ton and a half, the cable itself three tons, and the two-and-a-half-ton sphere contained no air and therefore was almost dead weight. As the winch ground to a halt and the dog was thrown into its cogwheel, the *Freedom* listed twenty degrees to starboard, the mast and boom creaked under the load, and all conversation on deck stopped. Beebe looked at Captain Sylvester standing by the winch, who shrugged. Then he glanced at Barton and saw the now familiar, wide-eyed expression of frightened defeat. "Bring it up. Slowly," Beebe ordered.

For an agonizing hour, the old *Arcturus* winch struggled under its load, as the cable popped and creaked. The tug's list drew the starboard rail closer and closer to the water, and on deck, there wasn't much to do but hope. As usual, Beebe tried to keep everyone's spirits up, but Barton fretted out loud until Tee-Van suggested that he get his movie equipment ready to record whatever was going to happen next. Barton and Fifilik set up a tripod aft against the wheelhouse, fixed the Kodak sixteen-millimeter camera to its mounting plate, and fired off some footage of the others standing with their hands on their hips looking down into the sea. Another large tugboat, the *Powerful,* steamed into view off the bow, prepared to get a line to the *Freedom* if her pumps failed or the weight of the sphere threatened to capsize her. Then Tee-Van shouted, "Here it comes," and Barton and Fifilik rushed to the rail and aimed the camera down where they saw first a flash of aquamarine and then the deep blue sphere that had disappeared into the depths an hour and forty minutes earlier.

The sea was calm, but the *Freedom*'s slight rolling motion was magnified in the swinging cable. The Bathysphere sloshed alongside the tug, and then it was out of the water and the men were steadying it in the air. Barton sent Fifilik back to a safer station in front of the wheelhouse and lent a hand on a guyline as they swung the heavy sphere to the deck. As the growling of the winch faded, everyone heard a sharp hissing sound and there was no doubt what had happened to the Bathysphere. Beebe looked over his shoulder at the rolling camera and pointed to a nail-thin stream of water shooting from the housing of the new window. He looked through one of the good windows, saw that the sphere was almost full of water, and motioned everyone to stand back. Barton handed him a hammer, then retreated with the others while Beebe tapped on the central wing nut in the hatch. It produced another powerful jet of water from the lower edge of the threaded bolt hole. The water inside had been driven into the sphere by the enormous pressure of the abyss, and the sealed steel chamber had locked it inside and carried it to the surface. Beebe tapped the wing nut again, and a fine mist enveloped him, accompanied by a higher-pitched whine. He motioned to Barton to move the camera to the side, and asked Tee-Van to stand on the other side of the sphere to help him. They turned the wing nut with their hands and listened as the high musical tone of the confined water gradually descended the scale a quarter note with each slight movement. Beebe and Tee-Van knew what was going to happen when the threads released the wing bolt so they stood well to the side of the hatch and fully extended their arms to turn it. Without warning, the heavy bolt was torn from their hands and fired across the deck like a cannonball to ricochet off the winch housing thirty feet away, causing a resounding report. For a few seconds, a solid cylinder of water gushed from the hatch. Then it dwindled to a trickle and everybody on deck began to breathe easily again.

Everyone's first thoughts after the explosion included the imagined consequences had Beebe and Barton been aboard the Bathysphere. The vision of performing the dreadful task of unsealing the wing bolt with their corpses inside flashed through everyone's mind. Beebe, though, recovered quickly, grabbed a thermometer from one of his field crates, and took the temperature of the water remaining in the sphere. Fifty-six degrees. He computed the temperature gradient from records of earlier dives and announced that most of the water probably had been forced into the sphere at a depth of about two thousand feet. There seemed to be no damage to the Bathysphere, except for the first coil of the threads of the wing bolt, which had been stripped when it was blown out of its housing. Tee-Van told Beebe they could file off the burred metal and still have plenty of thread left to seal the hole.

Beebe had Gloria Hollister write a news story about running for the harbor in the crippled *Freedom* and the Bathysphere returning full of water under immense pressure, and wired it to the *New York Times*. The article ran on the front page on September 14, and was picked up by dozens of other papers around the country and every station on the National Broadcasting Company network. The news that Beebe and Barton were going to sea on a tug that could sink at any time and descending into the sea in a steel sphere that might leak highlighted the risks of deep-ocean diving for millions of people, and helped build anticipation for the coming broadcast. Because no one was hurt, though, and the Bathysphere seemed to be in good condition except for the failure of the new window, the dramatic effect of a sphere full of pressurized water and Beebe's satisfying deductions of the depth at which it had entered made the test something of a success. Losing the third window was a disappointment, especially to Barton, but with the proven steel plug in place, the radio dive was not lost.

Eleven

ON THE AIR FROM BERMUDA

Courage is not the lack of fear. It is acting in spite of it.

Mark Twain

For another week, Beebe, Barton, the *Freedom*'s crew, and the radio men endured more bad weather but managed two unmanned tests in rough seas. They had replaced the third window with the old steel plug, but the Bathysphere again returned full of water and divers would have been killed had they been aboard. The sea had forced its way through the seal around the plug, as it had when the window was in place, with the same explosive result on deck. They sanded the seat in the housing and the steel collar, slathered on more white lead, replaced the plug, hammered the bolts down as tightly as humanly possible, and hoped for the best. Beebe tried to keep up the good cheer, knowing that the radio men were approaching their deadline of Sunday, September 22, beyond which they could no longer remain in Bermuda, and gloom set in at the St. George Hotel.

Finally, on Sunday, September 22, 1932, the radio dive was on. As the *Freedom* bucked a stiff chop on her way out to the cylinder, in the Black Rock neighborhood of Bridgeport, Connecticut, Henry and Mae Renaud called their children, Mae, Joan, and Raoul, into the parlor. Their home, like most of the others on the single block of Montgomery Street to the east of the foundry, was in a stacked triplex in which they lived on the ground floor with two other families, the Leibeys and the Massaitises, above them. The Renauds had returned an hour before from services at the Black Rock Congregational Church, where they had joined a few dozen of the non-Catholic immigrant French, English, and even a few Irish families in song and worship in the Pilgrim tradition, including among their prayers entreaties for the end of hard times, an

easy winter, and, from Raoul and a few others, the modest request that Max Carey's Dodgers beat Bill Terry's Giants that afternoon. (They did, 7 to 2.)

Henry Renaud wasn't particularly fond of the Black Rock Congo, as he called the little stone church, or the weekly ten-block walk to sing and pray, but he went along because Mae insisted that the great truths of life were to be found in the shadow of the cross and she wasn't about to deny them to her children. Henry, though, was more comfortable with the astonishing new discoveries about the earth on which he lived than with the conjuring of another world by a stern preacher. He had immigrated to America at the turn of the century, gotten a job on the killing line at the Morrell meatpacking plant in Bridgeport, and eventually become an accountant for the company. For almost two years he had been part of a skeleton crew working twenty hours a week at most, but in some of his newly found free time he had become an avid reader of travelogues and adventure stories, including those of William Beebe. He had also been following the recent newspaper stories about Beebe and his Bathysphere.

In the *Bridgeport Post*, on September 6:

BEEBE WILL TALK
FROM SEA FLOOR

Listeners throughout the country will hear the newest attempt of William Beebe, scientist-author, to plumb the depths of the sea in two broadcasts over combined networks of the National Broadcasting Company, on Sunday.

The preliminary preparations and the start of the descent at Nonsuch, ten miles off the coast of Bermuda, will be described by Ford Bond, NBC announcer, from the deck of the steamer *Ready* [*sic*] at 11:00 a.m. E.D.S.T. Beebe himself will describe his sensations as the steel bell in which he is encased is lowered into the sea.

At 2 p.m. E.D.S.T., when Beebe has announced that the steel ball has reached the expected depth of 2,640 feet beneath the ocean's surface, the broadcast will be resumed and Beebe will tell what he sees in the ocean depths. The technical arrangements of this unusual broadcast were directed by William Burke Miller, in charge of NBC's special broadcasts.

Henry Renaud settled his family in their regular chairs facing the mahogany console of an RCA Victor radio, which was half the size of their icebox. He told eleven-year-old Mae to turn it on, her privilege as the oldest child. She walked to the cabinet, reached for a large brown knob to the left of the tuning dial, glanced over her shoulder at her father, who nodded to let her know she was reaching for the right switch, and turned on the radio. Nothing happened. By now, though, a year after the marvelous instrument had come into their home, no one was alarmed by the silence, since Henry had explained that the tubes inside had to warm up before the radio would work. After five minutes, a steady hum took over the room, at which point Henry himself rose to perform the still-arcane task of tuning in the right station, WEAF New York. He found it quickly, right on time, and a resonant voice crackling through static joined the Renauds of Bridgeport.

Hello, New York and all points, this is the NBC group calling from Bermuda, Ford Bond speaking. Good afternoon ladies and gentlemen; we are speaking to you from the tropical waters off the coast of Bermuda, where the National Broadcasting Company has sent a staff group to bring to you an account of the New York Zoological Society's deep-sea exploration, under the direction of William Beebe.

The National Broadcasting Company engineers have set up micro-phones here aboard the Zoological Society's boat, S.S. Freedom, *located approximately seven miles due south of Nonsuch Island, of St. George's, Bermuda, from which the particulars of the Beebe exploration down one-half mile under the sea will be sent by shortwave to Bermuda, and thence through the courtesy of Imperial and International Communications to our New York studios. Then over the National Broadcasting Company networks through which you are receiving this feature.*

In Connecticut, Henry Renaud actually applauded. While Ford Bond's voice faded into static for a moment, he quickly told his wife and children how this miracle was being accomplished. He had read a story about the broadcast in the *Bridgeport Post* and for two weeks had been fascinated by the details. A radio transmitter with the call sign ZFB-1 was set up on the deck of the boat that would lower a hollow steel ball and its two occupants down into the ocean. From the boat, the signal would go through the air to radio station ZFB in Bermuda, then by an underwater telephone cable to Netcong, New Jersey, then also by telephone cable to hundreds of stations around America, which would send the signal through the air to radio receivers in homes like theirs. Not only that, he said, but the station in Bermuda was also sending the broadcast to Europe, so British and French families were listening as well. Ford Bond's voice again broke through the static of the shortwave signal from the deck of the tug.

This is a broadcast of a scientific undertaking. It is not a planned event to make a radio program. There are too many attendant dangers and the exploration is too important to forecast now just exactly what or how any-thing may happen. Dr. Beebe will attempt to descend to a depth five times greater than any other human being has ever been. He will literally enter a new world, from which we will bring you his voice telling of what he sees there. You, at your loudspeakers and we, here at the microphones are priv-ileged on-lookers, ex-officio members of the expedition as it were and in that capacity we shall stand close to where Dr. Beebe and his associates are working here aboard the S.S. Freedom *and hear the preparations being made for the descent. Later when he has attained a depth approximating one-half mile, we shall listen in on his conversation from the depths with his associate scientist, Miss Gloria Hollister, who is responsible for the sci-*

entific data which Dr. Beebe will gather from the depths and give to her
over the phone from which we shall pick up his voice from below. Miss
Jocelyn Crane is beside her to aid in jotting down the observations.

Dr. Beebe is now standing talking with his associate, Mr. Otis Barton
who will descend with him. I can hear their words from here . . .

On Bond's cue, McElrath swept over to the Bathysphere, where Beebe
and Barton were busy installing the oxygen tanks, checking the stuffing
box, and loading the rest of their equipment for the dive. The radio man
held the microphone near the hatch to record the sounds of metal on
metal as the tanks were clamped into place, and eavesdropped on Beebe
and Barton, who made awkward small talk about their chores. For the
first time in two years, the Bathysphere was being readied for a manned
descent. Some of the tasks were stale in their memories, so there was genuine concern in their voices. They commented on the weather, though
Beebe was careful to conceal the true extent of his trepidation about the
sea conditions, which were getting worse instead of better. Calling off
the dive at that point, however, was unthinkable. Two hours before,
Beebe had told McElrath he would attempt the descent. McElrath had
called New York and told them the broadcast was on, and the Sunday
music programs scheduled for 1:30 to 2:00 and 3:00 to 3:30 were canceled. With only two hours' notice, NBC stations were standing by in
New York, Hartford, Boston, Worcester, Providence, Schenectady, Buffalo, Detroit, Cleveland, Montreal, Toronto, Philadelphia, Baltimore,
Washington, Richmond, Pittsburgh, Ashville, Jacksonville, Miami,
Tampa, Nashville, Birmingham, New Orleans, Shreveport, Cincinnati,
Chicago, St. Paul, St. Louis, Kansas City, Davenport, Des Moines, Okla-
homa City, Tulsa, Denver, Salt Lake City, Butte, Seattle, Spokane, Portland, Ore., and Columbia, S.C. The British Broadcasting Company was
beaming the signal to southern England, Wales, and northern France.

A broadcast from the moon could not have drawn more of a radio
audience. On cue, McElrath dipped his microphone toward Hollister
and Tee-Van to capture the unrehearsed action of preparing to launch
the Bathysphere. They were acutely conscious of the imposing object in
McElrath's hand with the silver letters "NBC" mounted on a chrome
ring over a fist-sized microphone, and they spoke in stage voices and
leaned toward it. Hollister reviewed the emergency light signals that

could be triggered from the submerged sphere if the phone went out: one flash, okay; three flashes, pull us up immediately. Tee-Van said, "Right. Let's test it again." He reached inside the sphere and flipped the switch on the searchlight while Hollister watched the signal light blink on the instrument panel of the generator. "Looks good," she said, leaning into the mike. Ford Bond took over from there.

That was Gloria Hollister and John Tee-Van, both members of Dr. Beebe's expedition and on the staff of the New York Zoological Society. While some of the preparations are going forward for the descent let us as briefly as possible sketch to you a picture of the situation here, and tell you of some of the things we have observed in the several days which we have been on Dr. Beebe's boat during preparations for today's dive.

Dr. Beebe and his associates are so engrossed in the preparations for this important and dangerous scientific attempt and their attention is so completely absorbed in the details on which life itself depends, that we, a lay member of the expedition, will attempt to tell you something about the nature of it.

You of course know Dr. Beebe, celebrated for the colossal amount of material which he has added to the total of human knowledge. He, as an officer of the New York Zoological Society together with Otis Barton, [have] conceived this newest method of exploration and has carried the idea to this stage, where an exploration at a depth of one-half mile below the surface is deemed practicable, possible, and of great value to the scientific knowledge of this world in which we live. This attempt to reach two thousand, six hundred and forty feet under the sea, where pressures of over 6000 tons are exerted on the sphere in which he is traveling, is not to be taken lightly. Indeed, the amount of preparation necessary to ensure as far as possible the safety of the scientists, is the work of years. Also, the amount of thought necessary to make this diving sphere of practical use for observations under the surface has consumed the energies of Dr. Beebe and his associates for a long period of time. The question naturally arises as to how and what has and can be done to ensure both safety and scientific value. Dr. Beebe's Bathysphere seems to answer that problem.

. . . Inspection of all the vast details of completely closing the Bathysphere is going on now as we speak. Dr. Beebe is now minutely going over every opening, the two quartz windows, the door and its center open-

*ing where a huge wing-bolt will close the last opening, and the valve
through which the light lines and voice lines pass. Listen to him and his
associates . . .*

At 1:40, radio listeners heard the banter of Beebe and Barton crawling
through the hatch over Tee-Van's burlap sacks. Their voices entered
McElrath's microphone and flowed through the transmitter in Bermuda,
the transatlantic phone cable, and the NBC control room in New York,
and into radios in millions of homes, including the Renauds' in Bridgeport, where the children were fidgeting, a little bored by the announcer
describing something that simply was not part of their reality or imagination. Henry, though, was on the edge of his chair and his excitement was
contagious. He told his children that Beebe and Barton were minutes
away from being lowered into the ocean to a depth that was the same as
the half-mile walk from their house down Fairfield Avenue to Uncle
Nick's grocery store.

Ford Bond took over again.

*Remember, we mentioned no air lines. Dr. Beebe must take his own air
down with him. A small hollow tube would have no chance whatever of
passing air down to the Bathysphere. The terrific pressure would crush it.
Looking into the Bathysphere we see no suggestion of comfort—only the
hard concave black-painted steel—black to make vision better from the
inside and a searchlight mounted so its beam will pass through one of
the quartz windows, two oxygen tanks, two wire trays, one spread with
soda lime to absorb the expelled carbon monoxide, the other with calcium
chloride to take up body moisture.*

*The other equipment is simple—a telephone set with a breast transmit-
ter to leave the hands free, palm leaf fans to stir the air around them, and
the few switches needed to operate the phone and the lights. That then is
all of this daring projectile which will carry Dr. Beebe and his associate
Otis Barton down to the depths to study the deep sea creatures and also the
penetration of light at various depths.*

*Dr. Beebe has demonstrated the value of seeing the deep as the fish sees
it. He has trawled the depths to one and one-half miles but those fish
secured in deep sea nets are something very different down there in their
element instead of here in our light, pressured atmosphere. He has secured*

*temperatures at great depths, samples of the waters there, but now he will
go into that world himself, armored against its elements and report back to
the surface world what he sees.*

Bond's voice took on a sense of urgency now, much like that of a baseball announcer telling the crowd that the home team's best hitter is step-
ping up to the plate.

*. . . It seems that [the] minute inspections are complete . . . The huge
four-hundred pound door is being fastened over the manhole with ten
large bolts. The door has a circular metal gasket which fits into a shallow
groove. The joint is packed with a little white lead and the lugs hammered
down. Needless to say it is entirely water-proof. In the center of the door is
a wing-bolt plug which can be screwed in or out quickly.*

Beebe and Barton were inside, and McElrath stood next to Tee-Van
with his microphone as the blows of the hammer on the bolts and then
the wing nut rang through the air. Then, for a split second, only the
whistle of the wind made its way into the radio signal. McElrath and
Tee-Van backed away from the Bathysphere, leaving it alone on the foredeck, and Bond leaned down from his perch on the upper deck.

*The great Bathysphere in another aspect looks like a great blue ball,
painted blue to attract some of the great big fellows which inhabit the
deep. There is nothing like a ball for the even distribution of pressure.
That is why rain-drops are round and why Dr. Beebe's diving tank is a
hollow steel globe. Baits are also tied around it in various advantageous
positions to attract the deep's inhabitants.*

Ford Bond looked down at a stage direction on his typed script:
IMMEDIATELY AFTER THE DOOR IS HAMMERED SHUT THE ANNOUNCER
WILL AD LIB UNTIL BATHYSPHERE DISAPPEARS FROM SIGHT. But he was
speechless as the cable snapped taut and the sphere rose to his eye level
on the top deck, rotating slowly to the rhythm of the rolling tug. On its
first revolution, Bond saw the expressions of the men inside close to the
two windows, and was shocked to see a sternness akin to fear etched into
their faces. As the boom swung outboard and the windows rotated past

Bond again though, he noticed that Beebe and Barton were smiling, and then the Bathysphere was suddenly in the water. Bond struggled with his next words before regaining his composure.

Down . . . down . . . into the depths . . . two men in a hollow steel globe . . . and while you are listening to other programs for the next hour, the Bathysphere will be going down . . . down . . . down . . . down.

END OF FIRST BROADCAST

Twelve

VOICES FROM THE DEEP

As the Bathysphere settled again into the Atlantic, fading from deep blue to aquamarine to invisibility, Mae Renaud in Bridge-port snapped off the radio and the family arranged itself around their big pine-board kitchen table. Joan said grace and they dove into a mound of potato pancakes drenched in butter and Vermont maple syrup. Henry noticed that his wife and children seemed to be less impressed than he with the radio broadcast, so he rambled on at length about the mortal danger Beebe and Barton were enduring while his family and millions of others were enjoying their Sunday lunch. He told them about water pressure, the thin cable on which the lives of the divers depended, and the strange creatures of the deep ocean that no one had ever seen before. The children perked up, wide-eyed, when Henry said the men would be crushed to the size of a stack of pancakes if the sea broke through the windows, hatch, or wall of the Bathysphere, and his wife bowed her head and said a quick prayer that this would not happen.

Aboard the *Freedom,* Ford Bond and the radio men fiddled with their transmitter on the top deck and watched in awe as the Bathysphere vanished from sight, leaving only the evidence of the cable entering the ocean, slashing a foamy wake as the tug rolled in the building swell. Ahead of the *Freedom,* the *Powerful* again stood sentry against a suite of disasters, including the sinking of the *Freedom.* Since the first day aboard the leaking mother ship, the grim possibility that the tug would founder or sink when the Bathysphere was in the water had become more real than ever before, and the *Powerful* would be the last resort to snag the cable if the *Freedom* went down. Tee-Van directed the deck crew as they paid the Okonite line over the side, fastening it every hundred feet with rope ties, and threw water on the winch to cool it in its labor of lowering the weight of the Bathysphere and its cables. Hollister and Crane stood at their posts on the starboard rail with their notebooks open like hym-

nals, scratching down every word transmitted from the darkening world below. The *Freedom* was wallowing on the building swell and everyone was queasy, but so far no one had gotten seasick.

And in the netherworld of the Bathysphere, Beebe and Barton did their best to quell their fear and make methodical observations, but they realized immediately that the rough sea would make this dive wildly different from any they had made before. As they hung at two hundred feet while the rope tie was attached above, the motion was so violent that they had to devote every bit of their strength and attention to avoiding painful collisions with the oxygen tanks, the light box, the chemical trays, and the hard steel walls of the sphere. The wrenches for tightening the stuffing box, two screwdrivers, and their other tools collected at their feet in the bottom arc of the sphere, where the men pinned them with their seating cushions. The dim light flowing through the windows cast a moving pattern on the walls, emphasizing the jerky, swinging motion of the sphere, and Barton was quickly seasick. As they felt their descent resume with a downward lurch, his tender stomach let go of his breakfast, turning the inside of the Bathysphere into a reeking swamp of vomit. Beebe was wearing the telephone headset and as Barton heaved, he blurted up the line to Hollister, "Oh God, Otis. Not now."

Neither man wanted to abort the descent before the second radio broadcast, which was scheduled to begin at 3:00, but it was shaping up to be the longest hour of their lives. Beebe had laid his reputation on the line with his public declarations of goals of a half-mile descent and the discovery of bizarre new creatures, so nothing short of a life-threatening leak or fire would be cause enough to end the dive. Though Barton had his camera with him, it was obvious from their first frightening minutes in the water that shooting a movie was out of the question. He was miserably seasick, but like Beebe he wanted to continue the descent and make the broadcast. The sphere careened down through three hundred and four hundred feet, swinging through a pendulous arc of about thirty degrees and shuddering as the cable snapped taut and then slackened with each roll of the tug. Both men were bleeding from scrapes with equipment, and taking heavy blows that in the increasingly chilly confines of the Bathysphere knotted up at once like cramps.

Beebe phoned up no observations until they stopped at a depth of five hundred feet for a rope tie. Barton threw the light switch once, Hollister replied that the lamp on deck flashed, so the system was working, and Beebe finally reported two fish swimming by the window, and then a dozen pteropods that looked like silver balls in the distance. On the surface, the sun had broken through the overcast; though the red, orange, yellow, and green had disappeared from the spectrum, the water around the Bathysphere was shimmering with an ethereal blue glow. In the beam of his penlight, Barton saw that his photographic exposure meter still gave a reading of 1/30th of a second with a wide-open lens setting, so he snapped a picture of Beebe at the window. The oxygen system hissed comfortably and despite the vile odor of vomit in the sphere, neither man was having any trouble breathing. The stuffing box, hatch, windows, and newly seated steel plug hadn't leaked a drop, so Beebe told Hollister they would continue the descent and hope conditions improved. She relayed the message to Tee-Van at the winch, turned back to her green notebook, and listened to the shrill, strained tones in the Director's voice. On his end, Beebe tried to sound like it was business as usual, sending up his reports of salps, worms, copepods, fish, and flashes of bioluminescence when the searchlight was off. He called off the thermometer readings as the temperature inside the sphere fell through 76, then 74 degrees, told Hollister about the heavy condensation on the walls, and assured her that he and Barton were breathing easily. As the Bathysphere descended through one thousand feet, Beebe described blue-black water exploding with hundreds of flashing blue bursts in what he called "a light zone." At 1,425 feet, Hollister held the headset toward the tugboats so Beebe and Barton could hear their whistles saluting the new world record and, two minutes later, at 1,500 feet, the Bathysphere stuttered to a halt to wait for Ford Bond to introduce them to the radio audience.

Except for a dim blue whisper of brightness at the portholes, the men sat in absolute darkness, acutely aware of the chill of the abyss as the steel around them cooled. The wind above was picking up, the *Freedom* was rolling more than ever, and Beebe and Barton acrobatically braced themselves as the Bathysphere lurched and shuddered. They hung at 1,500 feet for two and a half minutes and used the time to take inventory of the critical elements of their craft. Barton felt around in the rubble beneath his legs, found bags of soda lime and calcium chloride, identified them with his penlight, and replaced the chemicals that were now part of a soup in the bottom of the sphere where a pint or two of water had pooled. Beebe carefully inspected the two windows, the steel plug, the hatch combing, and the stuffing box and found no seepage at all, so the little bit of water must have come from condensation when the calcium chloride tray was depleted and the moisture from their breaths was not being absorbed. The air itself was far from sweet, but quite breathable, though Barton noted that they seemed to be using more oxygen than on earlier dives, most likely because the stress of the violent descent was causing them to hyperventilate. Beebe and Barton made no small talk, but their ordeal had brought an unusual measure of ease between them. All in all, things were within tolerable limits as Ford Bond signed on the air.

Hello, New York. This is the NBC group calling from Bermuda.

Ladies and Gentlemen, for the past two hours the group sent into the tropical waters of Bermuda by the National Broadcasting Company has been watching a small steel cable slip over the side of the S.S. Freedom *seven miles due south of Nonsuch Island, off St. George. The New York Zoological Society's deep sea expedition under the direction of William Beebe, for the exploration of the deep, has reached a point far in the depths of the sea. Two hours ago, we brought you the preparations for the dive of the Bathysphere carrying Dr. Beebe and Otis Barton, his associate. Now Dr. Beebe is down below talking to and giving his observations to Miss*

Gloria Hollister here on deck. We shall now cut our microphones to listen as he passes up the data which he is gathering.

Aboard the *Freedom,* Ford Bond cued McElrath, who threw a switch to tap the broadcast directly into the telephone line linking Hollister and the Bathysphere. Millions of listeners in more than fifty other cities in America and Europe, including the Renauds in Bridgeport, stared at their radios. And 1,500 feet beneath the surface of the Atlantic, the divers felt the weightless sensation of their descent beginning again and Beebe started talking. For an hour, he had been a frightened man trying to survive, but now he was William Beebe, naturalist and author, speaking on the telephone to his assistant on the surface, Gloria Hollister, and a larger audience than he had ever had in his life.

Miss Hollister is on the telephone from the deck of the Ready *[sic]. She tells me we are now at 1,550 feet beneath the surface. The color of the water is the bluest black imaginable. No effect of penetration of sunlight. I am impressed with the tremendous importance of animal light in this zone.*

It is as black as Hades. A school of brilliantly illuminated jelly fish with pale green lights came within three feet of window. I have never seen such brilliant light.

Four or five eels, two or three feet long just went by. These may be Zonichthys. *A string of salpa just floated by. Thousands of lights, some in elongate patterns. Here come four lights in succession. Amazing amount of light which is pale blue to green but brilliant. It looks like surface phosphorescence but must be normal luminescence for this world.*

We are at 1,700 feet. It is pitch-dark inside and outside and Otis can get no reading with the exposure meter. The temperature in Bathysphere is 72°F.

Six fish coming nearer with double rows of lights. They might be Melanostomids. Dozens and dozens of salpa.

We are at 1,800 feet. The reading of our oxygen tank gauge shows that half of it is gone.

Silver hatchet fish, Argyropelecus, *going by. These are large and brilliant. Large salpa going by. A school of pteropods going by. These are round and coiled. Dozens and dozens of animals of varied and different shapes going by.*

A big fish just went by.

A school of fish just went by. I can see them by their own brilliance even when they are in the shaft of our light.

We are at 1,950 feet. The ocean becoming rougher and surface rolling increasing. Bathysphere rolling badly again. Cut lip on window ledge. Barton struck head hard on hatch. The sea is boiling with lights and I can make out jet black comets.

As many lights as ever. Otis running motion picture camera aiming it out the window.

We are at 2,000 feet. Five hatchet fish just passed, Argyropelecus, *two or three inches long. The biggest fish yet just went by. It is shaped like a barracuda. The greenest light seen yet. I can actually get full outlines of fish. Loads of little . . . I don't know what they are.*

We are at 2,100 feet. The Bathysphere is rolling badly. Biggest fish yet in the distance. Two in sight with their lights coming and going which are a pale creamy color.

Four big fish are going by, and hundreds of others. A fish is going by that is at least six inches long and of a deep shape. I can see upper and lower sides with the outlines visible from photophores whose direct glare was concealed from me. This fish turned slowly head-on and then all illumination vanished. More jelly fish than anything else. Many pteropods. Everything all lights and lighted up.

We must now conclude this broadcast. Otis Barton and I bid you farewell from a depth of 2,200 feet beneath the surface of the Atlantic Ocean off Bermuda.

Beebe's broadcast from the abyss lasted twenty-five minutes. At 3:29, on schedule Ford Bond signed off from Bermuda.

Far from in the depths, 2,200 feet to be exact, the National Broadcasting Company has brought to you the voice of Dr. William Beebe. Down there in his Bathysphere he has entered a new world hitherto unknown to man. He is at this minute still at that depth. The microphones of the National Broadcasting Company have carried you with him, as a privileged member of his expedition.

As he has of course been completely in his work we have had the privilege of participating in real scientific exploration. This today, as you have seen, was entirely scientific in its scope and so fraught with danger and full of importance that there was no certainty as to just what and how anything would happen. We have seen this great scientist actually at his work, have heard his reports as they will actually later be correlated and added to the sum total of human knowledge.

We wish to thank Dr. Beebe and his associates, the New York Zoological Society, the officials and people of the beautiful islands of Bermuda for the untiring hospitality and aid which they have shown to the NBC group. Mr. George McElrath, one of the executive engineers of the National Broadcasting Company, Mr. William Resides, and Mr. Edward Wilbur, Field Engineers of the Company and your announcer have received hearty support at every turn, to aid in bringing this broadcast to you. We hope that we have succeeded in making the picture both clear and enjoyable to you. Now, good afternoon. This is Ford Bond announcing.

Radios clicked off all over North America and Europe, and 2,200 feet below the surface of the Atlantic, William Beebe and Otis Barton listened to Bond's parting words through Hollister's phone set. As they hung for three minutes during the radio signoff, the pressure outside their dark chamber forced an inch and a half of the Okonite line through the stuffing box. The motion of the tug above was transmitting a rolling sine curve down the main cable and the Bathysphere was snapping and jerking more violently than ever before. Beebe and Barton were bruised and bleeding, and both had violent headaches. The stuffing box was succumbing to pressure, the broadcast was over, and though they were 440 feet short of a half mile, neither man felt there was any choice but to end the dive. Beebe ordered Hollister to begin the ascent.

Beebe continued his observations. At 2,100 feet, waiting for a rope tie to be cut off, he saw the lights of two big fish flash less than ten feet away. He turned on the searchlight and saw that the six-foot-long monsters ghosting past the windows were the size and shape of barracuda but with shorter jaws armed with fangs illuminated from inside their bodies. The fish kept their mouths wide open the whole time he watched them. A single line of bright, pale blue lights was strung along the sides of their bodies, and they had very large eyes. There were two long tentacles hanging down from their bodies, each tipped with a pair of separate luminous globes, the upper one reddish, the lower one blue. They also had vertical dorsal fins placed well back on their bodies, which meant they might be melastomids, or sea dragons. Beebe watched the fish for a few seconds, and then they were gone.

A half hour later, the Bathysphere broke the surface; the crew manhandled it to the rolling deck and had the hatch cover off in two minutes. Beebe and Barton emerged to wild cheers and applause from everyone on deck and the customary hooting of the whistles of the *Freedom* and the *Powerful,* buoyed enough by their success and survival that the two hours and two minutes of battering and terror vanished in celebration. After accepting congratulations, Beebe retired to the tiny radio room off Captain Sylvester's wheelhouse and composed a telegram to send to NBC and the Zoological Society in New York:

AT SEA ON TUG *FREEDOM,* FOUR FIFTEEN P.M. SEPT. 22. OTIS BARTON AND I HAVE JUST EMERGED FROM THE BATHYSPHERE AFTER REACHING A DEPTH OF 2200 FEET, PRACTICALLY A HALF MILE; 770 FEET DEEPER THAN OUR QUARTER MILE RECORD MADE TWO YEARS AGO. TELEPHONE COMMUNICATION PERFECT. NBC BROADCAST FROM 1:30–2:00 AND 3:00–3:30 P.M.

FROM 1700 FEET DOWN MY EYE REFUSED TO RECORD ANYTHING OTHER THAN ILLUMINATED MONSTERS. THE LARGEST FISH SEEN CLOSE TO WINDOW WERE 6 FEET LONG WITH BRIGHT LIGHTS. AT GREATEST DEPTH, THE WINDOW SHOWED ONLY THE BLACKEST OF DARKNESSES WITH 100S OF LIGHTS SHOWING ALL THE TIME.

JOHN TEE-VAN WAS IN CHARGE OF ALL DECK APPARATUS AND GLORIA HOLLISTER TOOK DOWN ALL NOTES SENT UP BY ME. WE SPENT TWO FULL HOURS IN THE SPHERE AND USED UP 40 GALLONS OF OXYGEN. THE SCIENTIFIC RESULTS ARE MOST SATISFACTORY. THIS VENTURE OF THE NEW YORK ZOOLOGICAL SOCIETY MAY BE CALLED A SUCCESS.

WILLIAM BEEBE

Thirteen

ANIMALS FAINTLY SEEN

They're dreadfully fond of beheading people here.

Lewis Carroll, *Alice in Wonderland*

B eebe knew that few moments in life are sweeter than those spent savoring the survival of an ordeal, so he laid on champagne and dancing to a ukulele band at the St. George Hotel for a farewell to Ford Bond and the radio men. Only Edwin Conklin joined them from the biological station, but the tables on the veranda were filled with Bermuda's social and government luminaries, who came to bask in the glow. Telegrams of congratulation trickled in all evening.

PROGRAM TODAY ONE OF THE MOST THRILLING IF NOT MOST THRILLING I HAVE EVER HEARD STOP CONGRATULATIONS AND HEARTIEST THANKS STOP PHILLIPS CARLIN, NBC NEW YORK

CARRYING ON YOUR WORK WITHOUT APPARENT REGARD TO RADIO AUDIENCE PROVED ONE OF THE MOST GRAPHIC BROAD-CASTS EVER PRESENTED STOP OUR SINCERE THANKS AND APPRECIATION TO YOU, MISS HOLLISTER, MR. BARTON, MR. TEE-VAN, AND OTHERS WHO HELPED BUT WERE NOT HEARD BY THE RADIO AUDIENCE STOP WILLIAM BURKE MILLER, DIRECTOR, SPECIAL BROADCASTS, NBC

GREATLY PLEASED WITH THE SUCCESS OF YOUR RECORD DIVE. CONGRATULATIONS STOP MADISON GRANT, PRESIDENT NEW YORK ZOOLOGICAL SOCIETY

Newspapers across the United States joined in the applause with stories on the historic radio dive, including this account by United Press International which ran on September 23.

BEEBE IN BALL
ON SEA'S FLOOR
RADIOS TO U.S.

SCIENTIST DESCRIBES STRANGE
FISH FLASHING LIGHT INTO
DARKNESS 2,200 FEET DOWN

NEW YORK, SEPT. 22 (U.P.) Huddled inside an airtight ball 2,200 feet below the surface of the ocean off subtropical Bermuda, William Beebe, explorer, broadcast his impressions to the United States today.

At a depth never before reached by a living human, Beebe and Otis Barton, inventor of the "bathysphere" in which the descent was made, peered through a tiny window at marvels of the deep, making com-

ments which echoed a split second later from radio sets in living rooms of Bridgeport, Portland and Peoria, Newark and Natchez.

The explorer was connected by telephone wires with the deck of the S.S. Freedom off Nonsuch Island, Bermuda. On that deck stood one of his assistants, Miss Gloria Hollister, chatting casually with the scientists in the depths below.

USED SHORT WAVE

Their questions and answers were short-waved by a National Broadcasting Company transmitter on the ship to Bermuda, and thence to New York.

"Rolling like the dickens" in the ocean depths, Beebe reported what he saw. At one moment it was "black as hades" outside—a second more and the iridescence of passing tropical fish rendered it "very brilliant."

It was only a moment later that it was "black as hades" again, but once more the descending sphere passed "some very brilliant lights."

"Now there are fish two or three feet away," Beebe reported. "I can make out their forms from their own light."

HEARD GRINDING CAMERA

At 2,000 feet the "bathysphere" was stopped for a time. The grinding of the camera could be heard plainly across the United States.

Fish were about them, but Beebe, an expert on ocean lore, had "never seen anything like them before." He saw "loads of little . . ." and the voice paused. He added:

"I don't know what they are."

For two weeks after the radio dive, a tropical low settled around Bermuda and brought miserable weather. The expedition hunkered down again and the revelry after the record-setting descent gave way to the routines of life at the biological station laboratories. While Hollister worked on the list of animals from the dives and Barton supervised the development of his film, Beebe compiled notes for his book and rushed to prepare what he hoped would be a sensational announcement of his discovery of a bizarre new deep-ocean species for the fall number of the *Bulletin of the New York Zoological Society.*

The business of identifying and naming newly discovered animals is as central to lives of field scientists as liturgy is to clerics, and the roots of the process are ancient. The system for organizing and naming living things in terms of their relationships with other living things began to evolve at the same time as the foundations of Western civilization themselves. More than two thousand years ago, Aristotle wrote about poking around in tide pools and classifying the colored shapes and blossoms of the seafloor. His most basic distinctions were "sensate" or "insensate," "animate" (which means "having a spirit") or "inanimate." He constructed elaborate charts describing the connections among animals in terms of their internal and external similarities, in essence classifying them. When the legendary Greek genius couldn't figure out whether something was animate or inanimate, such as one of the colorful and puzzling clumps of life we now call sponges, he just said they were "intermediates," or in between. Until the eighteenth century, classification was pretty much a matter of Aristotelian groupings of things that looked, sounded, felt, or tasted alike, or were somewhere in between. There was no single accepted system for organizing things into species or families of creatures, and no one was thinking about where animals came from because everyone presumed that all life had been generated spontaneously by God.

In the mid-1700s, a Swedish botanist named Carolus Linnaeus finally set up a method for naming living things that became standard and has been in use ever since. When Linnaeus was alive, about ten thousand animals had already been described, and he imposed an orderly filing system for them with species grouped within successively larger groups in a hierarchy based on very specific traits shared by all members of the group. For instance, a human being is of the kingdom Animalia; phylum Chordata; subphylum Vertebrata; class Mammalia; order Primates; family Hominidae; genus *Homo;* and species *sapiens.* When scientists talk about plants and animals, they usually refer to them just by their genus and species, such as *Homo sapiens.* This was an enormous breakthrough for scientific inquiry because it provided a means of describing and classifying animals that was universal among generations of scientists. With the Linnaean system, analysis and classification performed in the past informed current discoveries, and a rigorous system of peer review evolved to ensure that the foundation upon which future knowledge would be laid was sound. First, a scientist who discovers a new creature must propose and defend the identification of the species in one of many specialized journals that subject such discoveries to experts in the field of inquiry who must approve it for publication. After approval and publication, the description, which includes a species name that is entirely the choice of the discover, is accepted into the literature. Because classification is so important, transgressions are matters of enormous scandal in the community of scientists, mistakes invariably sully reputations, and fraudulent identification is tantamount to expulsion from the club.

Like most naturalists, Beebe was devoted to the elegant architecture of the Linnaean system. He had already contributed the accepted identifications and classifications of dozens of new fish from his deep-ocean trawls when he sat down at his desk in Bermuda to write a preliminary description of not only a new species, but a new genus of fish he had observed from the Bathysphere.

A NEW DEEP-SEA FISH

WILLIAM BEEBE

On the twentieth dive in the Bathysphere, at a depth of 2,100 feet, we saw two large, elongate, barracuda-shaped fish, which twice passed within eight feet of the windows, once partly through the beam of our electric light. These were at least six feet in length.

No direct lights were visible on the head, yet the rather large eye and the faint outline were distinct. There was a single row of strong, pale blue lights along the side, large and not far from twenty in number. The mouth, with strongly undershot jaw, and numerous fangs was illumined either by mucous or indirect internal lights along the branchiostegals.

The fish reminded me in general of barracudas, with deeper jaws open all the time. Posteriorly placed vertical fins were seen when they passed through the electric beam. There were two ventral tentacles, each tipped with a pair of separate, luminous bodies, the superior reddish, the lower one blue. These twitched and jerked along beneath the fish, one undoubtedly arising from a mental base, the other so far back that its origin must have been at the anal fin. Neither the stem of the tentacles nor paired fins were distinguishable.

I assume from the position of the vertical fins and the general facies, that the position of the fish must be somewhere near the Melanostomiatidae, but the single line of large, lateral photophores and the two ventral tentacles set it apart from any known species or genus.

The depth was 2100 feet, the date September 22nd, 1932, the position 32°17' No. Lat., 64°36' West Long., 5 miles southeast of Nonsuch Island, Bermuda.

Relying on this recognizable diagnosis, I propose for it the name of *Bathysphaera intacta,* the Untouchable Bathysphere Fish.

Helen Tee-Van drew the fish, illustrating the key points of Beebe's description, and he mailed the article and drawing to New York in time for the autumn publication of the *Bulletin of the New York Zoological Society.* Beebe had published dozens of descriptions in the *Bulletin* and the society's technical journal, *Zoologica,* but neither was considered to be in the top ranks of taxonomy because of inconsistencies in their peer review. Beebe had also published many articles and descriptions in *Nature, Science, Auk,* and *Copeia,* which were the guardians of taxonomic excellence, but out of loyalty to the society he decided that the formal proposal of a name for *Bathysphaera intacta,* which he called "the supreme discovery of the expedition," would go to the *Bulletin.* While he waited for the weather to lift, Beebe also prepared a paper he was to deliver before the National Academy of Science at its annual meeting on November 15, 1932, in which he would report the results of his work in Bermuda. In it, he mentions the sighting at 2,100 feet of "two large, elongate, barracuda-like fish." He does not use the name *Bathysphaera intacta* in his presentation.

By the first week of October, Beebe had finished his new description and the text of the academy paper. Finally, on the morning of October 8, he and Barton risked somewhat rough seas to dive again. They went only to one thousand feet because neither man wanted to risk a descent deeper than the harrowing radio dive in less than perfect conditions, deciding to concentrate on observation and filming instead of pushing their luck. This was dive number 21 for the Bathysphere, including manned and unmanned descents, and despite the leftover swell of the storm and a freshening wind, it went off with an efficiency that was almost routine. Two years earlier, they had reached just eight hundred feet as guinea pigs in an unproven craft, and though the pressure and equipment failures that might spell disaster were still with them, Beebe and Barton had become confident deep-ocean explorers.

For an hour and ten minutes, Beebe and Hollister ticked off their increasingly familiar list of animals, this time including a shark and pilot fish at one hundred feet, the usual schools of myctophids, jellyfish, and ctenophores, and coryphaena and puffer fish, some of which swam right up to the window to check out the lobster wired there as bait. At 875 feet, Beebe hollered up to Hollister that he had seen an explosion outside, a tremendous, blinding flash in the distance that illuminated the sea around the Bathysphere, but they had no idea what had caused it. He saw eels, silver hatchetfish, thick clouds of plankton flashing in the searchlight, but nothing really new. Like any competent naturalist, Beebe settled into the patterns and details of the world below, observing not only the unexpected but the ordinary with equal relish because he knew that understanding the neighborhood and the behaviors of the animals inhabiting it required familiarity.

Barton stayed busy, changing lenses, calling for more power from the generator to light his scenes, and loading film in his movie camera. Beebe told Hollister when Barton was shooting so they could compare the images on the film, if there were any, with what he could see out his own window. Barton wasn't sure that his cameras were capturing anything that would sell movie tickets, but he knew that he was the only human being who ever pointed a lens into the ocean's darkness one thousand feet deep and he was learning a lot. The earlier dives had seasoned him and transformed the Bathysphere into a reliable observation platform, and he thought that it was only a matter of time before something spectacular made its way onto his film.

For dive number 22 on the afternoon of the same day, Beebe's cranky sinuses were acting up again. He gave his place in the Bathysphere to Hollister, who was beside herself with joy. By then, the wind was licking the tops of the building swells, so they made the descent and return in just twenty-seven minutes, with Hollister calling up observations and Barton running his camera the whole time.

After the two one-thousand-foot descents, yet another hurricane passed fifty miles to the south of Bermuda, bringing winds of eighty miles per hour and frightening seas that rolled up the slope of the archipelago's sunken mother volcano and hammered the south shore. The expedition was down to a skeleton crew of four at the biological station—Beebe, Hollister, Tee-Van, and Crane—and Barton held down the fort at the hotel, which became the de facto headquarters during storms. The wild weather season of 1932 finally ended in late October when Beebe and Barton made three dives in the shallows to depths of 50, 250, and 290 feet to observe the near-shore life and further establish the Bathysphere as an instrument for that kind of observation as well as deep diving. The contour dives were as risky as descents in open water, because a collision with a coral head or the bottom could kill them as surely as a plunge in two miles of water. When they tested the technique in 1930, they had come within inches of being dashed against a reef, but even by diving in just a couple of hundred feet of water Beebe and Barton were pioneers. Barton fitted the Bathysphere with a new pair of fixed wooden rudders to stabilize it as the *Freedom* towed it slowly through the water, and hired a seaman skilled at depth sounding with a lead line who would be stationed on the tug's bow to warn of obstacles and sudden rises in the bottom terrain.

Everything Barton had filmed in the realm of darkness looked like a pan across a dark sky. There were no images of deep-sea creatures, but in the brightly lit shallows his movie camera finally recorded a parade of animals and a subterranean world that no one but he and Beebe had ever seen before. As Beebe called up his notes to Hollister, she could hear the camera whirring in the background. The assemblage of animals in the shallows contained many of the familiar fish and invertebrates he had seen while helmet diving, but never in such great numbers over so long a time. Beebe saw angelfish, sharks, grunts, eels, mullet, barracuda, yellowtail, wrasses, parrotfish, snappers, angelfish, sea fans, brain coral, sea cucumbers, starfish, vast fields of sea urchins, grasses, and in some places the curious absence of any life at all.

On their last dive of the year together, on October 25, Beebe and Barton spent an hour and a half at depths down to 290 feet, at one point missing a large brain coral and disaster by a few inches. In the shallows, the Bathysphere was a sweatbox and they kept their palm leaf fans moving constantly. They covered a lot of ground, enchanted by what they saw. Though mariners had been recording the topography of the seafloor for centuries with lead lines, no one had ever seen so much of it at one time. The variations fascinated Beebe. They coasted for ten minutes over a broad sandy plain that gave way to a patch of gravel and then fractured into deep canyons as the *Freedom* towed them away from land. Beyond the canyons, a steep, sandy slope took over. The bottom again turned barren, and for the rest of the dive, over sand ridges sculpted by current and tidal surge, Beebe's dominant simile and descriptive idiom was death:

"The bottom is death like." "Here and there old dead plume skeletons, tall and isolated, apparently dead for ages." "Dead desert, no fish." Life returned on the ascent, though, when a school of giant triggerfish escorted the Bathysphere to the surface as though chasing a fishing lure. Back on deck, Hollister took and recorded the divers' heart rates: *Beebe, pulse 90; Barton, pulse 98.*

Barton wanted to film the launching of the Bathysphere, so Beebe asked Tee-Van to make the final descent of the year with him just off Gurnet Rock. Barton and Fifilik filmed the hammering of bolts, Tee-Van's hand waving farewell through the wing-bolt hole, and the splash of the sphere into the Atlantic, just after noon. Beebe and Tee-Van reached fifty feet ten minutes later as the *Freedom* chugged slowly away from the rock. They were coasting along fifteen feet over the bottom, passing over scattered coral, when Beebe spotted huge masses of rock and coral dead ahead and screamed into the headset: "Pull us up as fast as possible."

Hollister barked at Sylvester, who reversed the winch, and twenty seconds later the Bathysphere was at twenty-eight feet, barely clearing a range of rocky crags rising from the bottom like stalagmites. It was the closest call yet, and Beebe and Tee-Van were so shaken that neither spoke for five minutes except to tell Hollister they were alive. Finally, the divers went back to work and for the next half hour observed fish, jellies, and the bottom contours in the poor visibility of the near-shore water. They were at helmet-diving depth, so there were no surprises until they passed over a big wire loop, which they found out later was the Transatlantic telephone cable beginning its journey from St. George Harbour. The dive lasted almost an hour, and then Beebe and Tee-Van were again safe on deck, breathing the ocean air.

On the *Freedom*'s final trip home as the mother ship for the Bathysphere, Barton and Fifilik took movies and photographs of the members of the expedition, who stood proudly with the flags of the Zoological Society and the Explorers Club and then mugged with the bait lobsters that were still alive after their descent. Once inside the calmer water of Castle Harbour, Barton asked Captain Sylvester to anchor off Tucker's Point and had the empty Bathysphere launched into four fathoms of water. He and Fifilik suited up in helmets to photograph the craft underwater, but gave up after a frustrating hour of camera breakdowns and leaks in the waterproof housing. The 1932 season was over.

Beebe had already decided to exhibit his craft at the Century of Progress Exhibition in Chicago, and three days after the last dive he put the Bathysphere into storage to await shipment to New York. Beebe made a short speech on the Darrell and Meyer wharf, thanking Captain Sylvester, the Bermuda Biological Station, and the people of Bermuda. And he promised to return.

The radio dive had been a smashing success, and though the film Barton shot during the deep dive showed nothing but a few flashes of light which really couldn't be distinguished from the usual nicks in the emulsion of his film, he and Fifilik had plenty of footage of deck operations, underwater film of the contour dives, and shots of the Bathysphere on the surface. It was a start, Barton thought, and he might very well be on his way to becoming the Frank Buck of the ocean.

Beebe, however, still listed his shortcomings in his New Year's diary entry. Though he concealed his self-doubt behind a facade of intelligence and Victorian grace, he couldn't shake the feeling that he really was a stuntman. The radio broadcast was all well and good, especially since he would capitalize on the publicity and with luck find the money for another deep-ocean expedition, but he had come up 440 feet short of a half mile and added very little to his reputation among his critics in New York and Bermuda.

INTERLUDE, 1933

*Presuming to describe and assign generic and species names for animals
faintly seen through the bathysphere windows . . . that were thought to
be fish is not merely improper but fraudulent, even contemptible.*

Dr. Carl Hubbs, ichthyologist

On May 25, 1933, William Beebe and Elswyth Thane boarded the
Twentieth Century Limited at Grand Central Station for the
overnight trip to Chicago and the opening ceremonies of the Century of
Progress International Exposition. The organizers of the biggest world's
fair in history expected millions of people to visit the 427-acre site on
Lake Michigan where the Windy City was celebrating its one hundredth
anniversary. According to its charter, the exposition would "demonstrate
to an international audience the nature and significance of scientific
discoveries, the methods of achieving them, and the changes which
their application has wrought in industry and in living conditions." The
plucky promoters of the fair had gotten their financing just weeks before
the stock market crash in the autumn of 1929, had persevered despite the
Depression, and were opening their doors to a weary world in the worst
year of economic distress. There was, however, a whiff of hope in the air
on the shores of Lake Michigan that spring. Franklin Roosevelt had
moved into the White House in March, bringing with him not only his
charismatic brand of leadership but a willingness to use the presidency to
force Congress, banks, and industry into economic reform. Not much in
the way of cold cash was filtering down into the pockets of the desperate
unemployed, but something was different and that had to be good.

In Chicago, Beebe and his Bathysphere shared top billing in the Hall
of Science with Auguste Piccard, who then held the altitude record for
his 51,783-foot ascent into the stratosphere. An attempt to break that

record in a new balloon named *Century of Progress,* piloted by Auguste Piccard's twin brother Jean-Félix, was among the most highly touted of the fair's extravaganzas. The balloon was the largest ever built, twenty stories high, with its billowing panels held together by a new material called Scotch Transparent Tape; it would be launched from Soldier Field in early summer with international press and radio coverage.

An astonishing 25 million people went to the great fair in Chicago between May and December 1933, and it actually made enough money to stay open another year. For Beebe, the publicity value of his association with the Piccards and the exposition was orders of magnitude beyond even that of his radio dive. And he was all too happy to be recognized by the general public as one of the world's bravest, most important explorers and to leave the simmering controversy about his scientific contributions in New York. Beebe had delivered his paper "A Preliminary Account of Deep Sea Dives in the Bathysphere with Especial Reference to One of 2,200 Feet" to the National Academy of Sciences on November 15, 1932. No one had ever reported firsthand observations of the quality of light, pressure, and temperatures in the depths, and the section of his presentation on the physical conditions to which a deep diver is exposed was sensational. Following classical scientific methodology, he cited the prediction of illumination theorists that sunlight would disappear entirely to a human observer at 1,700 feet, and confirmed the hypothesis with his direct observations.

Beebe's summary of the animals he saw in the abyss, however, was less warmly received. He devoted more than two-thirds of his paper to detailed lists of species from more than 1,300 net trawls and a record of sightings at hundred-foot intervals during his 2,200-foot descent, including *Bathysphaera intacta,* the six-foot-long fish he proposed for classification in the *Bulletin of the New York Zoological Society.* Within a month, he was hearing rumors that ichthyologists, including Carl Hubbs of the University of Michigan and John T. Nichols of the American Museum, were moving from polite skepticism to bilious attacks on his discoveries. He needed more deep descents to have any chance of confirming his identifications, but as he accepted the applause with Piccard in Chicago, Beebe wasn't at all sure that he would ever dive again no matter how much publicity he could generate.

Before Beebe loaded the Bathysphere on a flatcar for the trip to Chicago, he had loaned it to the American Museum for a display in the rotunda of their new building on Central Park. In a little more than three weeks between January 27 and February 21, more than 25,000 people had filed through the exhibit, a good crowd for Depression New York, and most of them had known about Beebe and his Bathysphere before they got there. They had peered inside the battered blue steel ball in which two men had gone down into the depths of the ocean and marveled at Else Bostelmann's paintings of creatures that defied imagination.

Otis Barton put in an appearance at the American Museum when the doors opened for the Bathysphere display. He stood with Beebe, Hollister, Tee-Van, and Crane for photographs, and left the next day for the Bahamas. By then, Barton considered the Bathysphere to be a vehicle for propelling him into the movie business, though the photographic results of the previous season were dismal. His Bermuda film showed the Bathysphere disappearing into the sea, the action on the *Freedom*'s deck, and a few dim images of the interior during dives, but contained nothing remotely sensational enough to draw crowds to the Mayfair Theater on Broadway.

Fame was bankable, though, and while Beebe's name led in all the headlines, Barton was getting enough attention to consider himself a genuine celebrity. His face was recognizable to millions of Americans, and Camel used it on a promotional card to sell cigarettes, with an endorsement from Barton: *I smoke as many Camels as I like. They don't give me jittery nerves. Camels have a swell taste—mild and yet with a rich, mellow flavor.*

The time was ripe to take a chance, Barton figured, and if he could just manage to make an underwater movie, his name alone would help sell tickets. So he pushed a big piece of what remained of his fortune into the pot and organized an expedition to the Caribbean with a crew of top Hollywood talent. His cameraman, Floyd Crosby, had won an Oscar for the South Pacific saga *Taboo*, was an accomplished diver, and had even worked briefly for Beebe in Guiana. To direct his movie, he hired Ilya Tolstoy, a nephew of the Russian novelist, who had turned to filmmaking after surviving the Revolution as a cavalry officer. Barton's cameraman and director were happy to be working at a time when the streets were full of filmmakers without jobs; but they also liked the idea of making a movie underwater in the Caribbean with a famous ocean explorer who still had a little money.

Twenty years earlier, George and Ernie Williamson had pioneered undersea motion pictures in the same warm coral shallows of the West Indies. The Williamsons were the sons of a Liverpool clipper ship captain who came to America to capitalize on the boom in grain and cattle trade in Norfolk, Virginia, in the last decade of the nineteenth century. During his long voyages around the capes, Captain Williamson was a tinkerer who invented practical devices such as a collapsible baby carriage and an electric ship's signaling lamp, and absurdities such as a way to play golf on the ceiling using balloons. Eventually, he left the sea to set up a shipfitting company in Newport News, and in 1908 built an underwater chamber for inspecting ships in shallow water without sending down a diver. The riveted steel observation cylinder itself wasn't anything new, but Williamson came up with a system of interlocking metal sleeves with canvas gussets fitted to a hole in the top of the chamber that extended up to the surface. The tube was three feet in diameter, big enough for a man to slide through while hanging on to rungs inside, and strong enough to withstand the pressure down to sixty feet. Captain Williamson worked on what he called his "hole in the sea" for a decade, but it never really caught on.

In 1913, five years after Williamson consigned his submarine tube to a scrap pile behind one of his shops, his sons fell under the spell of Thomas Edison's much more promising invention, the motion picture camera. At about the same time, George and Ernie came across a magazine story about underwater artist Zahr Pritchard and another about an Englishman who had built a bunker in the steep bank of his pond with a window through which he could observe beneath the surface. The next time the Williamson brothers were at home in Norfolk, they persuaded their father to drag out the submarine tube, loaded it on a barge, and took it out into Hampton Roads at the mouth of the Elizabeth River. They didn't have a movie camera—almost no one did—but they crouched in the observation chamber with a still camera and took pictures of seaweed, pilings, and fish. With Ernie inside, George swam down to the window and held up a copy of *Scientific American* for a photograph, figuring that a little self-promotion of their historic event wouldn't hurt.

The Williamsons had no idea that they weren't the first men to take photographs underwater. In 1893, Louis Boutan, a French zoologist, had lowered a view camera sealed with wax and mounted in a 400-pound frame to the bottom of a harbor to take a ten-minute exposure of himself standing in a diving suit. Boutan took hundreds more pictures, experimented with magnesium powder lighting, and wrote *La Photographie Sous Marin* to document his work. A few years later, Simon Lake took photos from inside his pioneer submarine, the *Argonaut,* and *National Geographic Magazine* published a sensational series of images taken from a glass-bottomed boat off Catalina Island, California.

For the Williamson brothers, though, a few snapshots were just the first step in their plan to make movies underwater. Armed with the photographs taken from their father's hole-in-the-water, they raised money to launch the Submarine Film Corporation to build and test a similar device for filming beneath the sea. Investors were charmed by their enthusiasm, the enormous publicity surrounding the debut of Edison's motion picture camera, and the mystery of the ocean. In a year the Williamsons had a new observation tube and chamber they named the Photosphere. They also bought a French-made Eclair camera, forty pounds of brass, iron, and steel with precision gears, sprockets, and a variable shutter that would enable them to hand-crank film through the gate at sixteen to twenty-four frames per second. At the end of March, they went to the Bahamas.

Two months later, George and Ernie Williamson were on a steamer back to New York with 20,000 feet of exposed movie film. They had shot coral reefs and fish, staged scenes in which one of them in a diving suit walked around on the bottom discovering "treasure," and filmed the bubbly plunges of local boys leaping from a dock. Their tour de force was a showdown between a diver and a shark, set up by weighing down the carcass of a horse to attract the sharks. The movie they eventually released from their first expedition was called *Terrors of the Deep,* which critics in Chicago, where it premiered, hailed as "something never viewed before by mankind." Three months later, the Williamson brothers were catapulted even further into moviemaking history when they joined Carl Laemmle's Universal Pictures to produce a film version of Jules Verne's best-selling novel, *Twenty Thousand Leagues Under the Sea.* Two years later, the first scripted, eight-reel, underwater epic opened to rave reviews and made a small fortune for the Williamsons, Universal, and their investors. "If the rest of the picture were discarded," wrote one critic, "the undersea scenes alone would be worth three times the price of admission."

Barton set off for the Bahamas to join the exclusive club of underwater moviemakers and to gamble what was left of his inheritance on a career that would hold his interest and maybe make a lot of money. In Nassau, he set his crew up in a hotel, chartered a weathered ketch named the *Tramp,* and went to work trying to stage battles among divers, fish, sharks, and anything else that looked scary. They filmed an octopus eating a lobster in a shallow pool, but that only reinforced Barton's feeling that it was going to take a bit more high drama to attract mainstream moviegoers. Crosby suggested that a battle between a beautiful girl and a man-eating shark was what they really needed, so Barton dispatched the handsome Tolstoy to Nassau on a recruiting trip. He returned with a half dozen bathing beauties and Crosby shot them swimming on the surface of a lagoon with a school of barracuda, but couldn't persuade them to put on helmets and dive. Barton sent Tolstoy back to Nassau, this time with instructions to look up a friend named Harold MacCracken who had told Barton he knew a terrific woman who was always interested in an adventure. This time, Tolstoy came back with Florence Goodman, a petite, photogenic brunette with long, slim legs that Barton thought would make her look like a mermaid underwater.

They almost lost Goodman on her first day. She was in fifteen feet of water when the valve in her helmet froze, the air stopped flowing, and in a panic she tore off the helmet, pushed off the bottom, and clawed her way to the surface. At that depth, the air in her lungs had expanded under pressure to one and a half times its normal volume. If Goodman had held her breath instead of obeying her training and exhaling, she could have been killed by a bubble of air forced from her lungs into her bloodstream, called an embolism, which was the bane of helmet divers. She was shaken, but still game, and the shooting went on. Barton's dream scene was a battle between Goodman and a shark, but they couldn't manage to capture and pen up a shark for the job. He settled for a sequence in which she stalked a barracuda with a .22 caliber rifle and shot the fish at close range, causing it to go berserk. For the rest of the Bahamas expedition, Barton and his crew photographed seascapes, coral heads rising from snowy white sandy bottoms, and glittering schools of reef fish. Crosby told him, though, that the interest of the moviegoing public grew in proportion to the size of the animals on the screen, so without shark attacks, the film would not be a hit.

Barton figured that the lack of sharks in Bahamian waters was his main problem, so he decided to go to Panama after reading in Michael Hedges's *Battles with Giant Fishes* that the waters on both sides of the isthmus were teeming with them. After hemorrhaging money on the Bahamas expedition, though, he sent Crosby and Tolstoy back to Hollywood, hired only a journeyman cameraman named Ted Dority, and picked up a drifter named Harold Davidson in Panama as a deckhand. In Balboa, Barton chartered a shallow-draft launch named the *Silver Spray*, motored across the Gulf of Panama to Pearl Island, anchored, and set out a baited line to test the waters for sharks. Just before dawn, the clatter of their fishing rope paying over the side woke everyone: they had their shark, a fourteen-foot female tiger thrashing in circles around the boat. Barton and Dority managed to bring the big shark alongside and lashed her to the bulwark to wait for morning light.

At dawn, Barton and Dority put on their diving gear and lowered the shark, still wrapped in a rope cocoon, to the bottom. While Dority set up his camera, Barton gingerly cut the shark free, secure in the controversial opinion that a shark would not actually attack a man. This one didn't attack Barton, who stood at its head and poked it with a spear while Dority rolled film, but the shot was quickly ruined because the shark went into a frenzy that stirred the water into an impenetrable cloud of silt and mud. Barton and Dority froze, standing on the bottom in thirty feet of water with a tiger shark they couldn't see for company. When the dust settled, the shark was still there, lying motionless on the bottom. There was no shark attack in the can that day.

Or on any other day for the three weeks Barton was in Panama in the summer of 1933. He came back to New York without his prized scene, but with a dramatic notion of himself single-handedly filming a shark attack fixed firmly in his imagination, which was almost as good.

> . . . the shark must be captured uninjured, that is, not hooked in either the throat or gills. Then this Houdini who can twist or bite his way out of wire loops and hemp ropes, must be held until morning, protected from attacks of other sharks. You then suspend the prospective movie actor by an invisible wire, letting him hang from the boom a few feet clear of the ocean floor. The point of attachment of the wire to the tough hide of the shark's back should be at the center of gravity so that the shark remains suspended on an even keel.
>
> . . . When all is ready the diver strikes the shark a sharp blow that

usually will cause him to open his mouth in what we called his roar and rush outward, taking a curved course to the surface where he thrashes around in fury. At last, the shark tires and begins to circle down to the low point. This is the moment for which the diver is waiting. He closes the electric switch of the camera and after he hears the camera rumble into action, he steps into the movie himself. He awaits the circling shark with spear poised to strike a defending blow. On comes the shark. The diver's spear hits his tender nose. The big mouth opens, showing rows of razor-sharp teeth. The shark slides past the diver, turned upward by the suspending wire. If the wire holds and the camera hasn't stopped, the chances are the picture will be good.

While Barton was chasing sharks, Beebe had returned from opening the Century of Progress Exposition and he and Gloria Hollister were also cruising in the Caribbean as the guests of Edwin Chance aboard his yacht, the *Antares*. Officially, they were on the Seventeenth Expedition of the Department of Tropical Research investigating the marine fauna off the West Indies, Pearl Islands, and the Mosquito Coast of Florida. Beebe drank in the reassuring atmosphere of fieldwork, friends, and discovery, and the zoological prizes of the five-week voyage were a small collection of fish, including a rare reticulated frog fish brought back alive and a mounted array of flying fish wings. The trip wasn't a scientific triumph, but in *Zoologica* that fall, he and Hollister jointly published "New Species of Fish from the West Indies," in which they identified three new species, naming a new kind of flying fish after the yacht, *Cypselurus antarei*, a new damselfish *Eupomacentrus rubridorsalis*, and a little goby after Edwin Chance, *Gobiosoma chancei*.

The social highlight of the expedition was, as usual, Beebe's birthday party, held at a hotel in Key West on the last night of the expedition. The celebration began with conch chowder and Chance's reading of a poem sent by one of Beebe's friends in New York, T. G. Aspenwall:

Beebe is my scientist, I may not sloth
He maketh me to lie down on pink coral snakes
And restoreth my faith in the British nation
Even though I walk in the Valley of the Shadow of
Colonic Creeks

I will follow no other scientist, for Willie is
   always with me
His machete and water glass they comfort me
He preparest a foaming beaker before me
In the presence of mine Antarei
And annointest my mouth with Eno
Till my * * * * * (deleted by a wise censor)
Surely the Tee-Vans and Hollisters will follow him
All the days of his life
And he will dwell in the house of
The New York Zoological Society
Department of Tropical Research
Forever.

Before Beebe left Key West, Chance again broke out his checkbook and underwrote half the expenses for an autumn expedition to Bermuda. Back in New York, Beebe mined his usual sources for the rest but came up empty, so he took care of the balance himself and, with Tee-Van, Hollister, and Crane, sailed aboard the *Queen of Bermuda* on August 17, 1933. Beebe and his staff stayed in a cottage at the biological station, but he and Wheeler were fed up with each other, so the main chore of the Eighteenth Expedition was the purchase and outfitting of a house near St. George as a laboratory and the official home of the New York Zoological Society in Bermuda. Beebe moved his headquarters from Nonsuch Island and christened the waterfront property on Ferry Reach "New Nonsuch." He packed all his specimens from the temporary quarters at the station to his own lab, and forged ahead with descriptions of deep-sea fish from his nets and deep-sea observations. Beebe was feeling the heat from his detractors, but his instinctive reaction was to work harder.

He and Tee-Van continued their work on a book on the shore fishes of Bermuda, and Beebe churned out papers for *Zoologica* on two new genera, *Macromastax* and *Photostylus,* and on four new species of scaleless fish, based on examination and dissection of specimens from his trawls. He also continued to work on a book on his deep-diving expeditions, writing chapters on the NBC radio dive and his observations of light at the end of the spectrum. The working title for the book was *Half Mile Down,* and although Beebe had been quoted as using the phrase in lectures and interviews, he knew he had come up four hundred feet short of that mark. Beebe, a skilled, talented writer, knew he had what theater people called a third act problem: he had developed an interesting story through multiple layers of adversity and success but was missing a climax.

In December, when the curtain fell on the first year of the Century of Progress Exposition, Beebe went to Chicago to fetch his famous craft, though he wasn't quite sure what he was going to do with it. From the outside, it still seemed solid, but inside, the rusted racks and oxygen tank clamps and noticeably clouded quartz windows made it look like a museum piece instead of a living tool for exploration. Beebe knew that if he was ever going to dive again, he had to raise enough money not only for operations off Bermuda, but also for a complete overhaul of the Bathysphere.

Part Three

MARVELOUS NETHER REGIONS

Fourteen

GROSVENOR AND
THE *GEOGRAPHIC*

*I am keen to repeat the dives, making them of three to four hours
duration, and expect to add many new discoveries to those made
in 1932. Next to going to Mars there seems nothing more
exciting and unpredictable.*

William Beebe, in a letter to Gilbert Grosvenor,
December 20, 1933

In the winter of 1888, thirty-three men converged on the Cosmos Club
on Lafayette Square across the street from the White House in Washington, D.C. They had come on foot, on horseback, and in carriages
on the damp, chilly night in response to an invitation from Gardiner
Greene Hubbard, a well-off lawyer who had financed the work of his
son-in-law, Alexander Graham Bell. Hubbard's letter proposed a meeting to "consider the advisability of organizing a society for the increase
and diffusion of geographical knowledge," and he had sent it to explorers, military officers, geologists, engineers, and inventors who shared an
abiding passion for describing the physical extremes of the planet on
which they lived. Among them were John Wesley Powell, who had led
the first thorough exploration of the Colorado River; naturalist and
physician C. Hart Mirriam, the chief of the ambitious U.S. Biological
Survey; George Kennan, who had crossed 5,714 miles of Arctic Siberia in
a sled pulled by horses, dogs, and reindeer; William Dahl, a pioneer
explorer of Alaska; meteorologist Edward Everett Hayden; and Adolphus Washington Greely of the U.S. Army, who had led an expedition
farther north into the Arctic than ever before and was stranded for three
years before his rescue by Commodore George W. Melville, who was also
at the meeting at the Cosmos Club.

By the end of the night, the thirty-three men had passed a resolution

forming a society organized "on as broad and liberal a basis in regard to qualifications for membership as is consistent with its own well-being and the dignity of the science it represents." In other words, just about anyone could become a member of the new National Geographic Society. Two weeks later, the same thirty-three men elected Hubbard as their first president. In his inaugural remarks, he reinforced that egalitarian concept. "By my election," he said, "you notify the public that the membership of our Society will not be confined to professional geographers, but will include that large number who, like myself, desire to promote special researches by others, and to diffuse the knowledge so gained, among men, so that we may all know more of the world upon which we live." To accomplish part of its mission of diffusing knowledge, the society began publishing a magazine in October 1889.

An unpaid board of editors, most of whom contributed articles, published the *National Geographic Magazine* whenever they had enough material for an issue. And despite their declared intention to reach out to general readers, most of their stories during the first year were distinctly technical. Geologist and anthropologist W. J. McGee contributed "The Classification of Geographic Forms by Genesis," on mountains and terrain, meteorologist Edward Everett Hayden used charts and technical language to describe the Blizzard of 1888, and Hubbard himself wrote "Africa, Its Past and Future," part of which was a diatribe against slavery. One critic said that the magazine was "dreadfully scientific, suitable for diffusing geographic knowledge among those who already have it and scaring off the rest."

In July 1890, though, Volume 1, No. 3, contained a very approachable story, "The Rivers and Valleys of Pennsylvania," inaugurating the magazine's coverage of the United States. The next issue carried a swashbuckling account of crossing Nicaragua to survey the route of a proposed transoceanic canal. In the summer of 1890, the society and the U.S. Geological Survey jointly sponsored an expedition to explore Mt. St. Elias, the highest point on the boundary of Alaska and Canada, and ten men braved icefalls, avalanches, and floods to bring back maps, charts, and photographs of one of the world's wildest places. Then, in its April 1891 issue, the *National Geographic* published "Summary of Reports on the Mt. St. Elias Expedition," written by geologist Israel C. Russell, who had led the party into the icy wilderness.

Russell's account of the expedition established another emerging part of the *Geographic* formula with its dramatic first-person narrative of the adventure, in which man's knowledge of the world is expanded by triumphing over nature and adversity.

> Darkness settled and rain fell in torrents, beating through our little tent. We rolled ourselves in blankets, determined to rest in spite of the storm. Avalanches, already numerous, became more frequent. A crash told of tons of ice and rock sliding down on the glacier. Another roar

was followed by another, another, and still another. It seemed as if pandemonium reigned on the mountains.

By 1897, *National Geographic Magazine* was an illustrated monthly, publishing high-adventure science and exploration with photographs; but it was still losing money. The following year, Hubbard died, Charles Bell became the society's president, and the magazine was $2,000 in debt. Not discouraged, Bell outlined a recovery plan for the board of directors in the summer of 1898.

> Geography is a fascinating subject, and it can be made interesting. Let's hire a promising young man to put some life into the magazine and promote the membership. I will pay his salary. Secondly, let's abandon our unsuccessful campaign to increase circulation by newsstand sales. Our journal should go to members, people who believe in our work and want to help.

And Bell knew two young men, either one of whom he thought would be perfect for the job of breathing life into the society's magazine. Gilbert Grosvenor and his brother Edwin were the identical twin sons of Amherst College professor Edwin A. Grosvenor, a frequent guest at the Bells' home in Washington. When Bell came up with the idea of a full-time editor at the *Geographic,* he diplomatically felt out both brothers for the job, mindful of the delicacy involved, especially because Gilbert—known as Bert—was courting his daughter, Elsie. Edwin told Bell he wanted to be a lawyer and wasn't interested, but Bert said yes.

On April 1, 1899, six months before William Beebe's first day on the job at the Bronx Zoo, a slim, compact young man named Gilbert Hovey Grosvenor reported to an office across the street from the Treasury Building in Washington, D.C., to become the assistant editor of the *National Geographic Magazine.* Little more than a cubbyhole, the room was furnished with two chairs and a small table competing for space with piles of boxes containing *National Geographic*s returned by newsstands. For ten years the magazine had been run by volunteer editors, and it was nearly bankrupt after trying to compete with *Harper's Weekly, McClure's, Munsey's,* the *Century,* and hundreds of other popular titles. Despite its financial condition and the absence of capital, Grosvenor could not have been happier. He had been teaching French, German, Latin, algebra, chemistry, public speaking, and debating, and he thought an editor's job would be easy by comparison. The president of the National Geographic Society, Alexander Graham Bell, put some pressure on Grosvenor, though, when he told him that the magazine had to continue to appear and prosper, or the mission of the founders and the society's 1,400 members would fail.

For the next thirty-five years, at the same time that Beebe was becoming a legend at the New York Zoological Society, exploring the jungles of South America, forging his career as an oceanographer, and writing books about his adventures, Gilbert Grosvenor was creating *National Geographic Magazine.* He agreed with Bell—who became his father-in-law with his marriage to Elsie Bell on October 23, 1900—that the attraction of a subscription to a magazine was secondary to becoming a member of a geographic society whose explorers, scientists, writers, and photographers traveled the world in their service. Within a month of taking the job, Grosvenor had launched a membership campaign from lists provided by Bell and other members of the society's board, and his own well-connected father. On the finest stationery with engraved letterhead, he personally wrote letters to prospects, always beginning with "I have the honor to inform you that you have been recommended for membership . . ." For a year, he sent out hundreds of these letters a month, wryly boasting to his father that complimenting a person on his nomination to the society was a surefire way to get him to send a check.

And it worked. John Hyde, a statistician at the Department of Agriculture, was the official editor of the magazine, serving without pay along with an editorial board including some of the society's founders. Hyde and most of the board rankled at what they considered undignified hucksterism, but by the end of Grosvenor's first year, membership—and subscriptions—had doubled. Realizing that most new readers of the *National Geographic* would renew their memberships only if he was able to deliver a unique, high-quality magazine, Grosvenor also went to work on an editorial formula he thought would set his publication apart from competitors. First, Grosvenor asked other magazine publishers for advice, among them S. S. McClure, whose *McClure's* magazine was selling 370,000 copies a month after only seven years in existence. McClure told Grosvenor that he should give up the idea of selling memberships as a way of selling magazines; that he should sell more advertisements; that he should move the offices to the center of publishing in New York; and that he should change the name to something simpler that didn't mention the word "geography," because people hated the subject in schools.

Grosvenor ignored every one of McClure's suggestions, and, armed with his modestly increasing sales figures and Bell's support, he prevailed over the objections of some of the board members and stayed his course. He also made a point of redefining "geography" in terms of its Greek root, *geōgraphia,* which means "a description of the world." "It thus becomes the most catholic of subjects," Grosvenor told the society's board, "universal in appeal, and embracing nations, people, plants, animals, birds, and fish. We will never lack for interesting subjects."

Grosvenor nourished his vision of a magazine serving a geographic society and its members and delivered on his promise to redefine geography. In 1908 another piece of the *National Geographic* formula fell into place when the society made its first cash grant of $1,000 to help Robert E. Peary reach the North Pole. Peary left New York aboard his ship, the *Roosevelt,* in July of that year, and after sailing to the edge of the earth's northern ice cap and driving 413 miles with dog teams, he took a noon sun sight that told him he had reached his goal. *"The Pole at last!!!"* he wrote in his diary. *"The prize of three centuries, my dream and ambition for 23 years, mine at last."* Among the flags Peary raised over the North Pole was the banner designed by Elsie Bell Grosvenor, with its blue, brown, and green stripes symbolizing the sky, earth, and ocean, with the words NATIONAL GEOGRAPHIC SOCIETY. The society's magazine alone published the authoritative account of Peary's heroic expedition, thereby sealing its reputation as the world's window on itself. Grosvenor never looked back. On behalf of its members, the society continued to organize and support expeditions to reveal, as promised, the nations, people, plants, animals, birds, fish, and other wonders of a world that was still largely unknown to its inhabitants. And Gilbert Grosvenor published the stories and photographs.

By December 1933, when Grosvenor received a letter from William Beebe asking for money to finance another season of deep-ocean exploration, the circulation of *National Geographic* magazine was hovering at just under a million. In the economic rubble of the Depression, almost 400,000 members had failed to renew since the stock market crashed in the autumn of 1929, but the magazine was in no danger of folding. The society was still writing checks for expeditions, recently including attempts to set new records for ascents into the sky, and Grosvenor was trying to get as much publicity as possible from their expeditions. As a condition of support from the society, most scientists, adventurers, and explorers agreed to write exclusive accounts for the magazine and mention the *National Geographic* prominently in interviews for newsreels, newspapers, and other magazines. The March 1933 *Geographic* carried "Ballooning in the Stratosphere: Two Balloon Ascents to Ten-Mile Altitudes Presage New Mode of Aerial Travel," by Auguste Piccard. This was the illustrated story of Piccard's flights in a gondola suspended beneath a hydrogen-filled balloon to altitudes of 51,775 feet on May 27, 1931, and to 53,152.8 feet on August 18, 1932. When Beebe met Piccard at the Century of Progress Exhibition, they posed with the altitude-record-setting balloon gondola and the depth-record-setting Bathysphere and shook hands for the cameras. The words *National Geographic* were clearly visible on Piccard's gondola, and "New York Zoological Society" in darker blue over the azure of the Bathysphere.

Beebe left Chicago with the germ of an idea for another season of deep-ocean exploration planted in his mind. If the National Geographic Society was supporting ascents into the stratosphere, they would probably pay at least some of the expenses of another season of diving, even during the Depression. He knew that Grosvenor, who was not only the editor of the magazine but, since 1903, the president of the society, would not be able to resist the symmetry of flying his flag at both the highest and lowest points ever reached by man.

Beebe had already published an account of his 1930 dive to 1,425 feet in *National Geographic,* his first story for the magazine, which had paid him $2,200, but Grosvenor had offered him no support for either that expedition or his return in 1932. "A Round Trip to Davy Jones' Locker," twenty-five pages of text with illustrations by Else Bostelmann and Helen Tee-Van and photographs by Jack Connery, appeared in the June 1931 issue, and generated a pulse of publicity for Beebe that would not be matched until the NBC radio dive. His first story in the *Geographic* had come at just the right time to attract what little money his regular donors could muster and had caught the attention of Edwin Chance, who became a major supporter of Beebe's work.

While he had been writing and editing "Davy Jones' Locker," Beebe had worked with Franklin Fisher, chief of illustrations at *National*

Geographic and Grosvenor's right-hand man. In November 1933, Beebe wrote Grosvenor a brief note offering his services and the Bathysphere to the society, and followed it up in a meeting with Fisher in New York. On December 12, Grosvenor opened the door to the possibility of sponsorship:

> As Mr. Fisher was in New York last week I asked him if he would be so good as to confer with you in reference to your letter to me dated November 15. Mr. Fisher states that you would like to have the National Geographic Society cooperate in your next explorations of the deep.
>
> This is just a note to say that the National Geographic Society will be pleased to receive and give serious consideration to any proposal that you wish to present. As the Society is now considering the disposal of such funds as are available at this time for research in 1934, I should appreciate your proposal as soon as possible.

A week later, on December 20, Beebe mailed his pitch to Grosvenor.

> My own financial plans for the coming year are still wholly unsettled, but I shall answer your letter of the 12th without further delay. Mr. Fisher told me that you might be interested in sponsoring my deep sea work with the Bathysphere in 1934.
>
> The actual cost of the 1932 expedition as far as the half mile dive [*sic*] was concerned was $1,600 for the chartering of the barge and wages of the crew. I am going to Bermuda with Mrs. Beebe the first week in January and will verify those estimates as applied to 1934.
>
> The total expenses of myself and my staff of three this year was $4,300 which I paid myself. This I cannot do the coming year, so at present I am putting forth every effort into raising the sum from some of our trustees.
>
> If you wish colored and other photographs of recent and living deep sea organisms it would be necessary to equip the tug *Gladisfen* with our regular trawling apparatus and pull nets for a month or so. I am not sure of this cost but think it would be under $2,000.
>
> The apparatus will of course be supplied by me. For example Bathysphere, rubber and steel cables, chemicals, oxygen, trawling winch, cable, etc., etc. When you realize the many thousands of dollars that have gone

into the production of these it seems a most fortunate thing that you in your magazine activities and I in my scientific life work can cooperate at such comparatively little expense. In addition to the actual apparatus there are many photographs, unpublished or used only in the [New York Zoological Society] *Bulletin* relating to the implosion and actual half mile dive [*sic*] of 1932 which would supplement another year's work.

The participation of the Zoological Society would be in name only but absolutely necessary as they pay the salaries of myself and staff. This relationship would be of the most direct and simplest character as I control my department completely and the society is concerned otherwise with only the zoo and aquarium.

This letter is necessarily tentative, until I see what my share can be and what your interest suggests. For peace of mind and relief of my regular annual worry I should infinitely prefer that you set aside a sum adequate for the entire expedition.

I am keen to repeat the dives, making them of three to four hours duration, and expect to add many new discoveries to those made in 1932. Owing to the risk and constant danger I should insist on being absolutely in command and having whoever you sent as a photographer being under my direction.

The dives could not be made before the calm period of June, but I want to go down in March if possible. The bathysphere is at the Chicago Fair subject to my recall. I am leaving New York Jan. 3rd and will be back on the 12th. Please let me have at least a tentative reply before the 3rd.

The holidays intervened, and Beebe's deadline came and went. On January 10, 1934, he wrote Grosvenor with more details on his budget for the coming summer. He had learned that the leaky old tug *Freedom* had been scrapped, and that he would have to charter both the *Ready* and the *Gladisfen* and their crews as he had in 1930, pushing up the cost. After consulting with experts on glass and oxygen systems, he knew that the Bathysphere would have to be shipped to the Watson-Stillman shop in New Jersey for new windows and an up-to-date air system before he could safely dive in it again. When he added the expenses of salaries for his staff, which he had pared down to Tee-Van, Hollister, Crane, and Bostelmann, Beebe came up with a figure of $12,000 for a month of diving. Beebe told Grosvenor he could provide $2,000, most of it from his own pocket, but he needed $10,000 from the National Geographic Society.

For another month, Beebe waited on tenterhooks for Grosvenor's reply, which finally came in a telegram on February 8.

YOU CAN COUNT ON NATIONAL GEOGRAPHIC SOCIETY GRANT OF TEN THOUSAND DOLLARS STOP WILL SEND CONFIRMATION BY LETTER EARLY NEXT WEEK STOP KEEP MATTER CONFIDEN-TIAL AS FIRST PUBLIC ANNOUNCEMENT SHOULD BE MADE FROM HERE STOP KIND REGARDS GILBERT GROSVENOR.

Grosvenor followed up on February 16 with a letter elaborating on his terms.

Owing to our economic limitations, $10,000 is the maximum possible grant we can make in support of your deep sea expedition.

In regard to the expedition it is understood that the work is to be carried out under the name "National Geographic Society—William Beebe Expedition," that the explorations will begin around June 1, 1934, and that the work will require from two weeks to a month depending upon weather conditions. It is to consist of perhaps three descents in the Bathysphere to approximately one-half mile in depth and that you will conduct a number of trawls to provide specimens for study and illustration.

It is agreed that you will furnish two articles about the explorations of this expedition for publication of 4,000 words each, with numerous good photographs and two sets of paintings done by Else Bostelmann or some other artist of equal ability, each to be published as a series of eight pages of illustrations in color. This material is to be delivered within a reasonable time after the completion of the explorations . . .

As to the use of the name of the New York Zoological Society in connection with this expedition, we are pleased to cooperate with this distinguished and efficient organization. You need have no anxiety on this matter.

You will also be free to publish any and all scientific material from the expedition in other publications.

Beebe had never sold a story or a book before he had written it, and Grosvenor's terms weighed on him from the moment he read the letter. He was also worried that Grosvenor's suggestion that a dive be made to a record depth of a half mile might be a demand of some kind. Beebe had already set successive depth records culminating in his 1932 dive to 2,200 feet, 440 feet short of a half mile, but he had not carried the flag of the National Geographic Society when he did it. Still, he wasn't about to turn down Grosvenor's offer. Beebe was going back to the abyss.

Fifteen

NEW EYES

*No water-soaked or icy snowballs are allowed. No honorable boy uses
them, and any one caught in the ungentlemanly act of throwing such
"soakers" should be forever ruled out of the game. No blows are allowed
to be struck by the hand or by anything but the regulation
snowball, and, of course, no kicking is permitted.*

Daniel Carter Beard, "Snowball Warfare,"
in *The American Boy's Handy Book*

William Beebe and Otis Barton hadn't seen each other since the day they shook hands for the cameras when the Bathysphere exhibit opened at the American Museum of Natural History in January 1933, and neither man had made any effort to keep in touch. For over a year, Beebe poured out stories about his deep-ocean exploration that would appear in *Harper's, McCall's,* the *Proceedings of the National Academy of Sciences, Zoologica,* and the *Royal Gazette and Colonist Daily,* and newspapers continued to report on his achievements. Beebe made a point of mentioning Barton and sometimes John Butler in some of these accounts and interviews, but only in passing. Barton received few requests for comments himself. On one of his visits to New York and in a fit of pique, Barton fired off a letter to the editor of the *New York Sun* to protest the paper's emphasis on Beebe and demand that he be interviewed. He also complained to friends at the American Museum that Beebe had an egotistical hunger for celebrity and would share his fame with no one. It wasn't so much that Barton wanted more credit for building the Bathysphere and making the dives, though he had done all that, but every line of ink was money in the bank if he hoped to make a living in the movie business.

Beebe was furious. Not only was Barton accusing him of hogging the limelight, but the insinuation that Beebe was making his descents only for their publicity value added sting to the barbs of critics who doubted their scientific value. The day after Beebe received the first half of the money from the National Geographic Society, though, he honored the promise he made when Barton had given him the Bathysphere after the 1930 descents and invited him to come along.

In March 1934, Barton was back in the Pearl Islands off Panama, still trying to film shark attacks. He and his cameraman, Ted Dority, had had no luck staging battles between a woman in a diving helmet, a shark, and a sawfish. They tried mooring a dead horse near a reef as bait, and sharks swarmed around the carcass, but their frenzy turned the shallow water into an impenetrable curtain of sand and silt. Barton spent a few frightening, blind minutes on the bottom before his crew dragged him to the surface. After the debacle with the horse, he and Dority checked into the Colonial Hotel in Panama City and began thinking about going boar hunting for a while. Then, on the morning of March 6, the bellboy delivered a telegram with Barton's breakfast.

LEAVE LATE APRIL FOR BERMUDA WITH BATHYSPHERE STOP
NATIONAL GEOGRAPHIC PAYING FOR OVERHAUL AND DIVING
STOP PLEASE ADVISE YOUR INTENTIONS STOP BEEBE

Barton asked the bellboy to wait while he scribbled his reply, telling Beebe that he would definitely be in Bermuda for the diving that summer. He and Dority booked passage on the next ship to Key West.

By the end of March, Barton was back in New York and the Bathysphere had returned to her birthplace amid the sparks of welders and cutting torches in the Watson-Stillman machine shop. Beebe supervised the monthlong overhaul with Tee-Van's dependable help, shuttled out to the shop several times a week, and kept as much distance between himself and Barton as possible. While Barton kept tabs on the refitting job and made occasional suggestions to Beebe, the simmering rift over publicity had pushed both men to the limits of their tolerance for each other. Though neither was willing to abandon the chance to dive again, they hid themselves behind walls of uncomfortable formality and the occasional telephone call. Barton set himself up in a new apartment on West Fifty-fourth Street in the heart of New York's cinema district, edited his film from Panama, waited, and worried. The movies and photographs of the creatures of the abyss had been a total bust so far, and Barton wasn't at all sure that he wanted to risk a half-mile dive to the physical limits of the Bathysphere and its cable. Especially if he wasn't going to get much credit for it.

Beebe was also thinking about taking the Bathysphere to the extremes of its design strength, but with none of Barton's trepidation, if his account of his reunion with his craft can be believed.

As the Bathysphere rested on her present bed of steel filings she seemed as staunch and sturdy as ever. I would willingly have scrambled inside and trusted her to carry me down and back safely to any depth I chose. But the doctors of mechanics, gathered in consultation, were more skeptical and they began to assemble what in human hospitals would be stethoscopes and sphygmometers [*sic*].

I peered in through the center window and it seemed clear as ever, and then I was startled to see several radiating lines as from a fracture. I rubbed the glass and a small spider ran over my hand and at my touch the strands of cobweb disappeared. I left the experts to their intricate examination, and motored back to the city.

Three days later, Beebe went back to Watson-Stillman and learned that a pressure test had cracked both old windows at nine hundred pounds per square inch, equivalent to a depth of about two thousand feet. The steel of the Bathysphere itself seemed to be fine, as strong as it was the day it came out of the foundry. The hatch had leaked, though, and its copper gasket and the brass wing bolt that had been damaged in the explosive decompression after the first test dive in 1932 would also have to be replaced. Beebe ordered three new windows from General Electric, where P. K. Devers of the quartz glass division offered him a 20 percent discount if his company was mentioned in stories and interviews. Beebe got the same deal from the A. D. Jones Optical Works in Cambridge, Massachusetts, for grinding the windows to perfect shape and clarity. The engineers at Watson-Stillman were worried that some undetectable flaw in the housing of the third window was the culprit in the disastrous leak on the test dive in 1932, so they advised Beebe to install only two of the new windows and plug the other hole with the steel plug as before. They also came up with a new way of sealing the window housings by drilling threaded holes in the top and bottom of their turrets. Each window was inserted into its frame with heavy paper washers front and back, the main bolts hand-tightened, and white lead injected with a pressure gun into the top hole, filling the channel cut into the face of the housing until it oozed out of the bottom hole. Then the threaded plug was inserted and tightened in the bottom hole, and more pressure was applied with the gun through the top hole, which was then plugged. Finally, the ten main bolts of the housing were hammered tight. The quartz glass was no stronger than that of the original windows, but they and the new seals stood up in a pressure test to four thousand feet.

From the moment Beebe opened the hatch of the Bathysphere for the first time after its return from Chicago, it was obvious that he was going to need a new oxygen system. Just two years after he trusted his life to an atmosphere created with guesswork valves, open trays of calcium chloride and soda lime, and two palm leaf fans, the rusty remains of that primitive system sent a chill through him when he looked inside. Beebe remembered the brutal headaches he and Barton had endured from too much oxygen, and knew that for the deeper descents he was planning he needed improvement in that system. He asked Tee-Van to find out what was new and better in the technology of creating artificial atmospheres, and by the end of March, C. E. Adams of the Air Reduction Company delivered his design. He offered it at cost to Tee-Van, who assured him that the name of his company would be mentioned in press releases and accounts of the expedition.

The oxygen tanks themselves were the same seventeen-inch-long, four-and-a-half-inch-diameter cylinders holding three hundred liters each, but mounted vertically behind the divers instead of horizontally at their sides. Recently perfected Air Reduction bleeder valves topped each of the tanks, capable of delivering oxygen to within 1 percent of the gauge reading. They were a marked improvement over the original single valve that had to be switched from tank to tank and gave readings on its gauge that could be as much as 50 percent off, depending upon the pressure inside the sphere. The inaccuracy of the old valve meant that they had to err on the side of too much oxygen. The new precision valve would allow them to set it for a flow of one instead of two liters per minute. Adams added a small calibrated barometer to the oxygen system that would simply hang on the wall as a backup to the gauges. In the unlikely event that a valve failed and oxygen was flowing too fast, the gas would raise the pressure inside the sphere and the barometer would let the divers know what was happening before it was too late.

The system Adams designed for removing carbon dioxide and moisture from the air was also a marked improvement over open trays and palm leaf fans. His invention consisted of four wire-bottomed brass trays, two containing sodium chloride and two containing soda lime. The removable trays fit tightly together in a stack between frames welded onto the wall of the sphere, under which was fitted an electrical blower that would circulate all the air in the Bathysphere through the trays of chemicals every minute and a half. The blower would draw its electricity from the same line that powered the searchlight. Tee-Van also installed a small automatic temperature and humidity recorder, recently invented by Julien P. Friez and Sons. This would monitor the amount of moisture produced by respiration; by watching the recorder, Beebe and Barton would know more precisely when the calcium chloride had been depleted.

The communication and power systems remained essentially the same, with the Okonite cable carrying four lines, which split into pairs for the telephone and electrical systems. Bell Laboratories gave Beebe new headsets, lighter and clearer than the old ones, and all they asked in return were the originals for their company museum. Tee-Van added a junction box to the electrical line to send power to both the 1,500-watt searchlight and the new air blower. Beebe and Barton ranked their attempts to shoot photographs and movie film in the abyss as a critical failure of their expeditions, and the brighter light with a more powerful generator pushing power down the line might solve the problem. A motion picture or even a still photograph of a real creature of the abyss would be a coup of immense proportions. Barton could imagine a sensational shot of a school of hatchetfish swimming straight at the audience from the screen at the Mayfair Theater, and Beebe knew that a single clear picture of *Bathysphaera intacta* or any of the other animals he hoped to see and describe on his coming dives could silence his critics once and for all.

While he was waiting for Watson-Stillman to finish refitting the Bathysphere, Beebe left Barton in charge and went to Bermuda to work on the first of the two articles he had promised Grosvenor. He had already received the full $10,000 and hated feeling indebted. The Nineteenth Expedition of the Department of Tropical Research included Beebe, Tee-Van, Hollister, Crane, artists Else Bostelmann and George Swanson, and two junior assistants, Perkins Bass and William Ramsey. With ninety pieces of luggage and equipment, they sailed aboard the *Monarch of Bermuda* on April 28 and arrived in St. George two days later. Elswyth Thane sailed with them, but spent only a week in Bermuda before continuing on to England.

Beebe and his Crowd settled into living quarters at the biological station, thanks to Edwin Conklin and despite the still-chilly presence of John Wheeler. Beebe opened up the laboratory at New Nonsuch, where he wrote in a windowed corner overlooking Castle Harbour. He enjoyed the writing because he believed that his words in a magazine with a million readers would inform and inspire other scientists, and he envisioned a future in which dozens of Bathyspheres would be exploring the abyss based on his work. Hollister and Crane checked facts, the names of species, and the chronology of events for the story called "Kingdom of the Undersea," a summary of the techniques of Bathysphere diving and an account of the 2,200-foot descent in 1932. Six weeks after they arrived in Bermuda, the article was finished and Tee-Van took it to the post office in St. George on June 16. Beebe included a letter to Franklin Fisher, who would be editing the story for the *National Geographic.*

> I am sending the first article. I have worried a lot about the best way to divide up the two. I have never before in my life taken money in advance for any writing and this is the last time. Ordinarily I can scribble an article, send it off and that ends it. But this being under obligation to write is very alien to my psychology. Owing to the difficulties I encountered writing an entire article based on a single dive . . . I would like to suggest that you consider publishing one, rather than two articles.

Three weeks later, Fisher responded, agreeing with Beebe that "Kingdom of the Undersea" was a little thin.

> Several of our editorial staff have read the manuscript and have been interested in the natural history described. While written in your usual excellent literary style, unfortunately in our opinion it does not meet our present need. Considering the matter from the viewpoint of your letter of June 16, we are inclined to agree with you that there is actually but one general article of supreme interest in your present work, therefore we shall modify the original plan and be satisfied with a single running account of your experiences on this expedition, with a survey of the results accomplished against the background of your previous experiences, including possibly some of the material contained in the manuscript now returned. If you agree with this, please provide the single story in the amount of 7,500 to 8,500 words.

Beebe breathed a sigh of relief, even though his literary debt to *National Geographic* was still hanging over his head.

> Thank you for your letter. I don't suppose any author was glad to get back a manuscript before, but I was very much relieved to know that I could do it in a single one. This will make it possible to do justice to the experience without having to build up a false second climax which does not exist.
>
> I will send the plates and the article in time for inclusion in the Janu-

ary issue and will try to justify your cordial cooperation and understanding of the conditions.

On May 23 in New York, Barton got a call from Stillman, who said
that the windows had been installed and tested and the Bathysphere was
ready for shipment. Barton told him to wait until he could get to New
Jersey to watch another pressure test for himself, and Stillman quickly
agreed, since he did not want to shoulder responsibility for the consequences if the glass failed during a dive. The week before, E. E. Free had
told Barton that even though the windows were new, they were no
stronger than the first ones. If the first ones were good enough, so were
the replacements, but the originals had never been taken beyond 2,200
feet. To Barton, Free had avoided the question, and that was enough to
deepen his fear. The next day, Barton took the train to Aldene and
watched as the shop crew fitted a new kind of test chamber over the windows with clamps and gaskets and ran the pressure up to 1,500 pounds
per square inch to simulate conditions at a depth of 3,500 feet. The windows held, but just as the gauge marked that depth, the noon whistle
blew and the two men who had been conducting the test went to lunch,
leaving Barton alone with the test chamber still attached to the window
housing. On an impulse, he simply closed the valve to stop the decompression, opened another, and ran the pressure up again, this time to two
thousand pounds. The glass held.

Even though the windows had been tested to over a half mile in the
shop, Barton wasn't at all sure he wanted to take the risk of diving again.
He had his life to lose, and not much to gain, since his photography in
the abyss was unlikely to produce results and Beebe was still reaping
most of the publicity benefits. Barton had lost touch with John Butler,
who was unemployed and had suffered a nervous breakdown, but he
tracked down an address and wrote to ask his architect's advice about
going deeper in the revolutionary craft they had built together. The two
men had fallen out because Butler thought he had not received enough
credit for his work on the Bathysphere, so Barton wasn't at all sure that
Butler would reply even if he received the note.

Once more we are contemplating a dive in the bathysphere. This
time I believe we will reach ½ mile.
Before I decide whether or not to take part in this attempt I should

like your opinion as to weather [*sic*] the windows will stand this pressure. We can of course make a test to 3,000 ft. or more first. Dr. E. E. Free has advised me against going.

I am sorry for your current circumstances. I did the best I could for you in the matter of publicity in connection with the bathysphere. Dr. Beebe got control of the affair and I have no more say.

A week later, Barton slit open an envelope delivered by messenger and read Butler's reply.

My dear Mr. Barton,

Your letter was a surprise to me, as my last correspondence 2 yrs. ago when I sent you some information you requested was never acknowledged.

Re. my opinion as to whether or not you should take part in the attempt to descend to ½ mile, it is my opinion that you should go to the experts who have been handling the bell since I turned it over to you, completely equipped, 4 years ago. I lived up to the letter, on my part, of our gentleman's agreement.

Butler

Sixteen

BLOTCHES OF LIGHT
ON A BLACK FIELD

. . . for whatever we lose (like a you or a me)
it's always ourselves we find in the sea.

E. E. Cummings,
"maggie and milly and molly and may"

John Butler's curt letter didn't ease Barton's doubts about joining
Beebe's attempt to descend to a half mile, but in the end he decided
he couldn't afford to back out of the expedition. The odds were good
that the windows would hold, and there was always the chance that he
would get lucky with his cameras. More important, he wasn't about to let
Beebe set a new world depth record without him. So on the morning of
July 3, 1934, Barton arrived at the Furness Line Pier at the foot of West
Fifty-fifth Street to watch the Bathysphere being loaded unceremoni-
ously into the hold of the *Queen of Bermuda.* The *Queen* was painted the
same elegant shade of silver gray as the trousers of a formal morning suit,
with brilliant white topsides, and an interior fitted out to carry nothing
but first-class passengers on the New York–Bermuda run. Though the
Depression hadn't loosened its grip on most Americans, there were still
enough well-heeled travelers with the $65 one-way fare for the two-night
passage or the $150 "Honeymoon Fare" for the round trip on the luxuri-
ous 580-foot liner to make her pay. As the Bathysphere was lifted from a
lightering barge by the ship's cargo boom, Barton and Dority captured
the moment on film from the dock under the liner's bow, tracking the
sphere as it rose from the barge like a dark blue balloon and disappeared
over the rail. Then they packed up their camera and tripod, walked to
the passenger gangway, and boarded themselves.

Two days later, the scene played out in reverse as Beebe, Tee-Van,
Swanson, Barton, Dority, and a good crowd of locals stood on the wharf

in Hamilton and watched the Bathysphere rise from the hold of the *Queen of Bermuda,* revolve slowly against a crystal-clear blue sky, and settle onto the deck of the *Powerful.* The tug left the harbor and negotiated the shoals on the fifteen-mile trip to St. George, passing over the familiar coral wonderland and along the nearby coast, which was splashed with a riot of sea lavender, goldenrod, oleander, and primrose. Her passengers, however, were oblivious to the pleasures of the view. The bristling tension between the two men whose destinies were tied together by the little blue ball on the *Powerful*'s stern was as undeniable as a bad odor in a parlor. To Beebe, Barton had become little more than an annoying passenger who complained publicly that he wasn't getting enough credit for the Bathysphere descents. He acknowledged Barton's stroke of genius in bringing the Bathysphere to life, but he had little use for the man, who seemed to him to be lazy and given to mood swings from mania to gloom that made everyone around him uncomfortable. And while Barton still thought Beebe was a great adventurer and naturalist, he couldn't stand his superior attitude and really did suspect him of hogging the glory of their expeditions into the abyss.

Beebe couldn't take the tension for long. He pulled Barton away from the others and onto the bow of the tug, where they stood face to face, Barton with his arms folded on his chest, Beebe with one hand in the

pocket of his khaki shorts poking the air with the forefinger of the other as he did most of the talking. After ten minutes, they shook hands and walked aft to join the others who were standing around the Bathysphere. The bitter, distracting aroma of contempt was gone. Beebe made a rare note about Barton in his diary that night: "Had a run-in with Otis over his silly letters about my hogging the publicity, after which he was quite decent."

Though the rift between them never really healed, Beebe and Barton were pragmatic enough to realize that they still needed each other. For the next month they worked shoulder to shoulder fitting out the Bathysphere and her mother ship the *Ready.* During their first two seasons, they had felt their way down to 2,200 feet, but now they intended to push themselves and their equipment to the extremes, and the preparations took on a distinctly different air. Beebe and Barton were the only veteran Bathysphere divers on earth, and more than anything else, their experience meant that they knew how many things could go wrong during a dive. To have enough time in the abyss for breakthroughs in their observation and photography, and to set a new depth record, everything had to work perfectly. Their motto for the season, "Three hours and a half mile," became part of their repartee and was always the last toast of an evening on the veranda at the St. George Hotel.

The realities of meeting their goals translated into long days at the Darrell and Meyer wharf for Beebe, Barton, Crane, Hollister, Tee-Van, and the crew of laborers and deckhands hired for the season. It was a reunion of men, women, boats, and machinery already proven and celebrated, and the bonds of shared adventure invigorated everyone. The redoubtable *Ready* had spent the past two years hauling fuel oil and freight among the islands. Although her deck was a jumble of scrap and one of her two boilers was blown, she was still level on her lines. Beefy, mustachioed, and loud, Captain Jimmy Sylvester was back, too, by then one of the proudest local fans of the American explorers. Since the cranky old *Gladisfen* was undergoing an overhaul and would not be in service until mid-August, he would use the *Powerful* to tow the *Ready.* Beebe and Barton deferred to Sylvester in all matters to do with the tug and its barge, including his insistence on rebuilding the *Ready*'s second boiler and remounting the seven-ton *Arcturus* winch on her deck with a new, more heavily braced housing. One boiler probably would have been enough, and the winch in its old housing had worked just fine, but

Sylvester carried the caution of a true sea captain and never hesitated to sink any idea that might compromise safety.

While Sylvester and his crew worked on the barge, Beebe had the schooner *Taifun* brought alongside and the Bathysphere was hoisted over to the deck of the old three-master, where he pulled out the protective wooden plugs over the windows and hatch and gave the sphere a new paint job. It had been white the first year, dark blue the second, and now he would paint it sky blue, his theory being that a certain amount of attraction from a brighter color might be a good thing. Beebe hired a local sign painter to apply the white lettering on the side of his craft.

NEW YORK
ZOOLOGICAL SOCIETY

BATHYSPHERE

NATIONAL
GEOGRAPHIC
SOCIETY

Its name on the Bathysphere was only part of the publicity campaign the National Geographic Society was building around Beebe's descents into the abyss. Grosvenor had backed away from the day-to-day management of the society's participation in the expedition, and one of his vice presidents, John Oliver LaGorce, was handling the details. LaGorce was also in charge of another Geographic Society expedition, the stratosphere ascent of Major William Kepner, Major Albert Stevens, and Captain Orvil Anderson, who were set to ride in a sealed gondola similar to Piccard's under a three-million-cubic-foot balloon named the *Explorer*. The altitude record was thirteen miles, achieved by Russian balloonists in January 1934, and Kepner, Stevens, and Anderson would attempt to reach fifteen miles, or 79,200 feet, on July 28. Through Franklin Fisher, who was in close touch with Beebe, LaGorce asked his deep diver to send a good-luck telegram to his balloonists.

SUGGEST YOU SEND TO US RADIO GOOD WISHES FOR STEVENS
AND KEPNER STRATOSPHERE FLIGHT ABOUT READY TO HOP UP

THE PUBLICITY STOP GOOD FOR YOU BOTH STOP WE WILL
RELEASE STOP FISHER

Beebe dutifully sent his telegram, but he could not resist taking the
opportunity to suggest subtly that exploration was about scientific discovery, not setting records.

HEARTIEST GOOD WISHES TO STEVENS AND KEPNER FOR SPLEN-
DID OBSERVATIONS IN STRATOSPHERE FLIGHT AND HAPPY LAND-
ING STOP BEEBE

Good wishes, unfortunately, didn't do the trick for Kepner, Stevens,
and Anderson. They lifted off before dawn from a field in North Dakota
under a billowing balloon big enough to hold an eleven-story hotel. The
men in their spherical gondola, with the words NATIONAL GEOGRAPHIC
SOCIETY and ARMY AIR CORPS stenciled on its side, were ascending
through 60,000 feet when disaster struck. Through a quartz-glass porthole in the top of the gondola, the men watched in horror as the balloon
started to come apart. A half hour later, they were at 20,000 feet, falling
at a mile a minute under the tattered balloon, and the men abandoned
ship. They struggled to get through the small hatch wearing their para-
chutes, and Kepner was the last to jump at three hundred feet, seconds
before the gondola slammed into the prairie.

E. John Long, a Geographic Society public relations man for the
Explorer mission, put an upbeat spin on the expedition in the society's
official press release:

> We consider the Stratosphere Flight a success in that the three gallant
> gentlemen descended safely and we are further delighted to find that
> much of the scientific data recorded on photographic film throughout
> the flight was not destroyed on impact, especially important being the
> results of the cosmic ray detector and the work of the spectrograph.

The day after Long wrapped up that job, he sailed for Bermuda.
Grosvenor wanted all information on the Bathysphere expedition to
come from the society, which was fine with Beebe, except that there was
a lot of interest in his work and he was having trouble saying no to

reporters who came around to ask questions. Long would send cables, grant interviews, show reporters around, and handle anything and everything having to do with getting the story out to the world. Long was too young to be an authority figure and not an underling, so he became Beebe's new best friend with a quick wit and an easy way with the others, especially Jocelyn Crane, on whom he developed a crush the moment he met her. Long was part of everybody's crowd. Jimmy Sylvester liked him because Long grew up sailing and had a mariner's instinct for moving around a boat with all the proper deference toward the captain and crew. Even Barton liked him, because Long had applauded loudly the night Barton showed his twenty-minute film of the 1932 diving season at the theater in Hamilton.

On August 6, the members of the expedition, its press agent, the tug, the barge, and their crews assembled on the wharf, ready to go to sea. The Bathysphere was in position on the *Ready's* bow, looking like a showroom automobile in her new paint, fully equipped and ready to go with the improved atmosphere system, recording barometer, diamond-bright new windows, and a searchlight that could throw up to 1,500 watts into the abyssal darkness for photography. Barton, passionate about the new searchlight, had made a hurried trip back to New York for a larger generator that could deliver a variable power load, so he and Beebe would have to endure the full heat of the lamp only when actually shooting film. At 250 watts, the heat from the bulb was bearable, but 1,500 watts turned the tiny cabin of the Bathysphere into a sweatbox even with the chill of the abyss outside. The telephone and electrical line, laid out in great loops along the port-side rail, had also been overhauled, with six hundred feet of new Okonite cable heat-spliced to replace a section that had corroded. The *Arcturus* winch, as high as a man and half the deck wide, was freshly greased and plumbed into the two steam boilers. The seven-eighths-inch Roebling cable was rigged through a heavy steel sheave forty feet out on the bow, back to the stout wooden mast, through a second sheave, up to the boom, through a third sheave, and down to the stainless steel clevis that connected it to the lifting lug on the Bathysphere.

The weather was perfect, the sea as calm as a pond, and everyone had had enough of anticipation. Even so, Sylvester talked Beebe into a final full-dress rehearsal at the dock. Half of the sixteen-man deck crew had been part of the earlier expeditions, but the others had not been drilled on the tricky business of lifting the Bathysphere, lowering it into the water, bringing it alongside close enough to clamp the Okonite line to the lifting cable, keeping steam on the winch, and managing the entire process in reverse. "Three hours and half a mile" was not only a motto but a reminder that no matter what had gone before, the coming descents would bring the divers and the topside crew to the very edge of their abilities and concentration. Beebe was preoccupied with his visions of triumphant observations of the creatures of the abyss, and Barton with photographing them; but Captain Sylvester's obsession was imagining the worst things that could happen and preparing for them.

With the *Ready* still moored at the Darrell and Meyer wharf, they went through the choreography of launching the Bathysphere into St. George Harbour. For the first time in two years, Beebe and Barton squeezed in together and tried to get comfortable. In the dim light of the cramped chamber, any residue of ill-feelings between them morphed into something close to soldierly comradeship. Outside, Sylvester roared his orders to the crew, setting them at their stations near the winch and boilers and on the bulwark to pay out the power cable. Hollister took her position on the starboard rail with her headset and notebook, Crane stood near the winch to monitor the depth and keep track of time with a stopwatch, and Tee-Van performed the familiar chore of closing the hatch and sealing the center hole with the wing bolt. For the test dive, he only used four bolts to dog down the hatch cover, since they were only planning to dunk the sphere into the water. Tee-Van signaled Sylvester, the winch came to life with a rasping growl and the hiss of vented steam, and the Bathysphere rose from the deck. Inside, Beebe again mused that it was odd that every trip into the abyss begins with a twenty-foot ascent into the air. He watched the deck and the crew spin below him through the crystal-clear windows. He and Barton were exhilarated, all their doubts and anxiety banished by the thrill of being back in the Bathysphere, but the strained expressions Beebe saw on the faces below reminded him how hard it was to be left behind on a deck awash with fear.

Thirty seconds after they were in the water, one of the consequences of carelessness brought Beebe and Barton face to face with the blunt truths about the audacity of descending into the abyss. At a depth of five feet, Beebe was crouching at the center window watching a school of fry flash in the sun around the Bathysphere when he felt a chill around his ankles. As he reached down to investigate and felt his hand plunge into eight inches of water, Barton saw that the sea was pouring in from both sides of the hatch and screamed at Hollister to pull them up. In seconds they were again swinging in the air. Though the incident quickly became part of the joking banter of the expedition, no one could ignore the fact that relaxing any precaution, even for an instant, could be fatal. Beebe and Barton had already witnessed the horror of the Bathysphere full of pressurized water after the unmanned test dive two years before, and now they had experienced the moment when that same sea was actually bursting through the hatch.

They made a second test dive that day to twenty-two feet with all the hatch bolts tightened. Beebe and Barton sat quietly in the Bathysphere on the bottom of St. George Harbour while Sylvester and Tee-Van coached the crew through their assignments. Everything worked perfectly on deck and below except for the stuffing box. The inner nut was biting into the power line when Barton tightened it and leaking quite a bit even in the shallow water. Back on deck, they inspected the Okonite cable and found that its outer sheath of rubber was brittle and crumbling under the pressure of the stuffing box nut so it couldn't be tightened enough to hold its seal. The other end of the cable, though, seemed firm and pliant, so they spent the rest of the day reversing and rewiring the phone and electric line. While Tee-Van supervised the crew working on the Okonite cable, Barton and Dority started the job of rigging their camera with an automatic timer in its rack over the center window for the deep-water test of the empty Bathysphere they thought would come the next day. By the time the cable job was done, though, the sun was low in a sky gathering dark clouds from the west, the barometer was falling, and Beebe announced that there would be no diving the next day. Barton and Dority put away their tools and left for the hotel and the cocktail hour, figuring they would have plenty of time the next day to finish installing and calibrating their camera.

The next morning, Barton's phone rang at seven and he heard Beebe's cheerful voice telling him that the weather had turned out just fine, the test dive was on, and the *Powerful* and the *Ready* were sailing immediately. Beebe told Barton to meet him at the wharf at eight so they could motor out in the *Skink* and rendezvous with the tug and barge at sea around ten. Barton realized that unless he was aboard the *Ready* for the ride to deep water though, he wouldn't have time to finish setting up his camera, so he woke Dority, and still in their pajamas, they raced to the waterfront with two cases of equipment, arriving seconds late as the *Powerful*'s tow rope tightened and the barge eased away from the dock. Sylvester shouted that he couldn't turn back without running aground, so Barton and Dority jumped into a canoe tied up nearby, paddled furiously to catch up with the tug, which was moving slowly through the entrance channel, and were hauled aboard the *Ready* by the cheering crew.

After two hours of hot work inside the Bathysphere, Barton and Dority were standing in sweat-soaked pajamas when Beebe, Hollister, Crane, Tee-Van, Ramsey, Bass, and Long, dressed in crisp tropical whites, came alongside in the *Skink*. Barton didn't care how ridiculous he looked, though, because his equipment was ready, and, apart from seeing if the Bathysphere could return whole and dry from three thousand feet, the only mission of the test dive was to film what was down there. During the manned dives to come, Beebe's observations would take priority at the center window, and although Barton could aim his camera and push the button once in a while, photography would play second fiddle. For the test dive, however, Barton's new Bolex camera was in the center window. It was equipped with an automatic timer and hung from a removable wire rack on the ceiling arc that positioned the lens three inches from the glass. The searchlight was fixed in its permanent housing on the side of the sphere, aimed through the right window. Barton and Dority had the camera loaded to shoot four hundred feet of supersensitive panchromatic sixteen-millimeter film at a speed of sixteen frames per second for a total run time of six minutes.

At 11:30, Barton reached into the Bathysphere and set the timer to trigger the camera at 1:00, when the Bathysphere would reach three thousand feet. Tee-Van and Bass sealed the hatch and wing bolt and gave the signal to Sylvester at the winch, and three minutes later, the only sign of the blue steel ball was a ring of foam around the cable whistling through the choppy surface of the sea. On deck, the crew performed well enough to keep Sylvester from barking, as they tended the boom, paid out the heavy power cable, fastened it with rope ties to the lifting cable, and monitored the health of the winch. The weather meandered from bright sun and almost flat calm to driving rain in local squalls, which were irritating but didn't disturb the surface of the sea enough to force Beebe to call off the test. An hour and twenty minutes after the Bathysphere hit the water, Crane yelled "Three thousand feet" from her post at the cable gauge. On Sylvester's order the winch ground to a halt and a crewman clanged down a ratchet to hold the drum steady. The sun was back out. In the sudden silence the only sounds were the thrumming vibration of the cable and the creaking of the boom until Beebe said, "Otis. You're on."

Barton waited two minutes and when his watch read 1:00 straight up, he twisted the rheostat on the generator to push more electricity down the line to the searchlight. The engine labored under the load, a sure sign that a half mile beneath the *Ready*'s keel, the lamp was burning and throwing its weird white light into the eternal darkness. If the timer had worked, if the camera was rolling, and if the windows hadn't fogged from the sudden blast of heat from the lamp into the cold interior of the Bathysphere, Barton was shooting a movie in a netherworld where no one had ever gone. He left the light on for four minutes, then turned it off to test his camera and film on the living lights of the deep-sea creatures in the darkness, and, after two more minutes, told Jimmy Sylvester that he was done. The familiar hissing growl of the winch washed over the deck again, the cable crackled under the strain as it wound onto the drum, and the Bathysphere began its slow ascent toward sunlight.

There were no battles between bizarre creatures never before seen by man on Barton's film, nothing that would bring shrieks from a cringing theater audience, no shocking pictures of the sort he expected would elevate him to the ranks of cinema celebrity. Most of what flickered through the gate of the projector were smudges and blurry images moving past a lens that was out of focus, but the film did show, as Barton would remember for the rest of his life, "a bonita-shaped fish that, stunned by the dazzling ray from the window of the Bathysphere, paused, turned and sank slowly back into the blackness of the ocean. Almost the entire side of the fish registered on the film as a sheet of flame . . . It did not appear to be one of the fiery dragons, but whatever it may have been, it was the first and only deep-sea fish to be photographed in its native depths."

Seventeen

WORLD RECORDS

*The sea is a world of enchantment, far surpassing any described in
the "Arabian Nights" or fairy tales; a world teeming with life
so strange that some of it we can scarcely believe to be real.*

Daniel Carter Beard, *The American Boy's Handy Book*

After the unmanned camera dive to three thousand feet, the expedition was weathered in for three days. Beebe and Barton used the time to test their air system and check every inch of the power cable for breaks in the outer covering. They could not erase the memory of the dive in 1932 when the telephone line went dead at 250 feet and a sense of utter isolation fell on them like a shroud, and they did not want it to happen again. Beebe had himself sealed up alone in the Bathysphere on deck for two hours on a drizzly afternoon, chatted on the phone with Hollister about the birds he saw through the windows, watched the gauge on the oxygen tank, and kept an eye on the barometer. He found out that the humidity went up slightly when the searchlight was on, that it went up quickly when the blower was turned off, and that the arrangement of the chemical trays with the blower on the bottom allowed moisture to drip down into the wiring. There was plenty of room over the stack of trays, so he had the blower rewelded to the top after the test. It seemed to work just as well, with no chance of an electrical short. Beebe decided to carry a pair of palm leaf fans to back up the new system, just in case.

The morning of August 11 dawned bright and windless, "a grand day," Beebe declared, and by 9:30 the *Powerful* and the *Ready* were riding easily on a long southerly swell six and a half miles southeast of Nonsuch Island at 32°14'40" North, 64°35'40" West. Three reporters from New York newspapers, Long, Dority, and John Knudsen, a photographer from the

Zoological Society, stood on the bow forward of the loops of Okonite
cable, where Tee-Van told them they would be out of the way. As Beebe
was taking a last look around before climbing into the Bathysphere, he
overheard Long saying something to the reporters about attempting to
set a depth record and shot the publicist a chilly glare. After Beebe and
Barton were sealed inside, Long kept using the phrase "world record,"
but in his typically facile way he made a joke out of it. As the Bathysphere
swung free of the deck and hung twenty feet in the air with the National
Geographic Society flag flapping on the cable, Long said, "And that,
ladies and gentlemen, is a new world altitude record for a big steel ball
with two men in it and no balloon," which got a laugh from everyone,
including Beebe when Hollister relayed the wisecrack through the phone
line. Record setting was a tender subject, and though Grosvenor had
mentioned a depth of "approximately one half mile" during negotia-
tions, Beebe had carefully avoided promising any record or depth goal as
a condition of the deal. Of course everybody knew that anything over
2,200 feet would be a new world's record and that a half mile was their
acknowledged goal, so Long's witty banter was finally a way to admit that
there was an elephant standing in the corner that no one had previously
been willing to acknowledge.

At 9:41, the Bathysphere slipped into the Atlantic, and although Beebe and Barton had seen the transformation from the golden yellow world to a green one many times, it startled them once more. As they hung twenty feet down, the crew swung the boom closer to the *Ready's* side so they could tie on the cable clamps. Beebe snapped into his observational trance, describing the puckered surface from below and the sponges, shells, and sea grass on the barge's bottom. The hum of the blower made the Bathysphere seem safer and more alive than its primitive ancestor two years before, and the constant circulation had a noticeable effect on the sweetness of the air. When Hollister asked if they were ready to descend, Beebe and Barton had never been more comfortable telling her yes.

The colors vanished one by one. The sun was bright and high in the sky, so its light penetrated to the full distance each color could travel before the friction and density of the seawater removed it from the spectrum, giving Beebe a good record of its transformation into total darkness. Since he had first observed the full deterioration of the visible spectrum on his dive to 2,200 feet in 1932, Beebe had read more about the physics of light. In 1666, Isaac Newton passed sunlight through a prism and showed that it contains six colors always ordered in the same sequence: violet, blue, green, yellow, orange, and red. What Newton didn't know, but Beebe did, is that the colors and their infinite gradations exist because they are created by electromagnetic waves oscillating at different frequencies. The reason colors occur in their specific positions on the visible spectrum is that their wavelengths are different and fixed. The wavelength of violet is longer than blue, blue longer than green, and on up to the wavelength of red, which is the shortest and oscillates at the highest frequency. Shades of color occur because the boundaries between them are gradual and there are an infinite number of wavelengths. The combination of all the colors produces white, the absence of all colors, black. Beebe also knew that visible light is a small sliver in the very middle of the much wider electromagnetic spectrum, with oscillations ranging from very short waves on one end creating high-frequency gamma rays, to very long waves on the other end creating low-frequency radio waves. And finally, Beebe's observations of the dwindling of the spectrum as he descended into the abyss in 1934 were informed by a puzzling new way of thinking about the electromagnetic spectrum called quantum mechanics. For reasons that were understood by only a handful of people at the time, he could look out the window of the Bathysphere and imagine that the waves of light also existed as massless particles called photons.

Actually observing the behavior of waves and particles in the electromagnetic spectrum was critical to proving what theoretical physicists were predicting about their dual nature and the probability of their appearance in either form. Auguste Piccard had carried a gamma ray detector to measure and record high-energy waves on his balloon ascents into the stratosphere, where the rays are unblocked by the atmosphere and therefore more common and undisturbed. After talking with Piccard and later Grosvenor, Beebe hoped that descending into the living laboratory of the sea, where light is naturally transformed, would offer him a chance to make a contribution to another great scientific adventure. He searched for an electric spectrometer, an instrument that precisely measures the wavelengths of the spectrum, but the smallest one he could find was too big to fit through the Bathysphere's hatch, and he was forced to abandon his plan. Beebe would do the best he could with what he had, though, so he was using an improved version of Bostelmann's original painted spectroscope to confirm the depths at which the colors and their waves and photons vanished to the human eye.

Beebe concentrated on calling out the gradations in the sunlight to Hollister while also ticking off the life-forms moving past his window. Quickly, red was gone as though it had never existed. At 250 feet, flickers of silver dominated the twilight of water outside the Bathysphere, as copepods and fry reflected the flash of the sun off their dancing bodies.

Beebe's face was six inches from the center window, and Barton took quick peeks when he wasn't watching the stuffing box or the oxygen flow gauge. At three hundred feet, the remnants of yellow dissolved into green, and quickly the green was gone, too. The shift from bluish green to greenish blue was imperceptible, but at four hundred feet Beebe transmitted the new description to Hollister and fell into a reverie about a lovely colony of siphonophores looking like spun glass ornaments in the aquamarine glow. These colonial jellies are strands of specialized individuals, each performing separate functions such as flotation, swimming, stinging, feeding, and breeding, but all joined by a common food canal. Beebe had seen dozens of siphonophores crushed to oozing mush in his nets, with all their crystalline loops and tendrils lost in a tangled knot, but had witnessed their astonishing beauty only a few times in the depths. No one but he and Barton, looking over his shoulder, had ever seen them whole and alive.

Passing through five hundred feet, Barton flicked the lamp on at low power, and its beam was a faintly visible white streak in the diffuse, monochromatic blue water. The heat of the bulb in its tin housing brought an instant warmth to the interior of the Bathysphere. Beebe remained fixed at the window and as the Bathysphere continued its descent into darkness, he chanted his observations to Hollister. Squid, silvery brown. Many little copepods, a dust cloud of tiny creatures, water dark rich blue. Walls very cold, humidity 45 percent. Big single light, thousands of copepods, now about twenty pteropods, the long tubular kind. Many strings of salplike animals. Four-inch fish with six bluish white lights along their sides. Worms—no, they are round mouths, light-colored ones. Beebe was breathless from talking, but he had entered the realm of ecstasy that had first seized him when he was a teenager naming the birds around his house in New Jersey.

With a thousand feet of the sea above them and five hundred pounds pressing on every square inch of the Bathysphere, Beebe and Barton took stock. They had been in the water for thirty-two minutes, the stuffing box and the door were bone dry, the noise of the blower was not interfering with telephone conversation, and the humidity was so low that Beebe did not need a kerchief over his nose and mouth to prevent the window from fogging even though the steel of the sphere was becoming very cold. The surface of the sea above was calm, so the sphere was rock steady at the end of the cable. Beebe tried again to name the water color, blackish blue or maybe dark gray-blue. Barton switched on the bulb, its faint white beam illuminated the darkness, and motes of microscopic animals flashed like dust in the only light they had ever known. Beebe knew that sunlight was still present and would be for another five hundred feet, but tracking the deterioration of color was useless from that point on. He could concentrate on the creatures of the darkness. In the dim light overflowing from the lamp housing, he looked at Barton and said, "Okay?" Barton nodded.

At eleven hundred feet, the sea was filled with flashes of light. When Barton switched on the beam again, Beebe was rewarded with the greatest display of life in the abyss he had ever seen. He ordered a hold at that depth for five minutes, watched, and described what he saw out the window. Three fishes going past, appeared out of the darkness. Twelve flashes going on and off. Large dark body passing, squid or fish? Three full-sized *Argyropelecus* swimming upright and together, larval fish, a whole string of luminescence spread out like lace. Many copepods and sagitta, all very active in beam. Nothing very large. Jellyfish very luminous from food. First eel, lighted up from beam. *Serrivomer?* Eight inches long, did not go into beam. Something large up close to edge of light, a very deep fish. Four-inch leptocephalus, not deep, swimming obliquely upward. Small siphonophore and *pyrosoma,* one foot long with no light. Beebe told Barton to switch off the light, took a breath, and watched the sparkling lights in the darkness.

Remembering his disappointment at the transcripts of oohs, aahs, and wordless exclamations from earlier dives, Beebe concentrated intently on his observations, allowing himself only brief rests. He had coached Hollister to ask him questions to focus his mind as it raced to keep up with what he was seeing, and she peppered him with "How big?" "What color is the light?" and "How many?" He knew that the better the record of his sightings, the more defensible the extraordinary would be later when he classified his discoveries. Beebe wasn't about to be ambushed by Carl Hubbs or anyone else who claimed he was making up animals that didn't exist. At 1,300 feet, with the faint beam again illuminating the sea for a few feet in front of the window, he was rewarded with the spectacular sight of a three-inch anglerfish with a pale lemon-colored light on a slender tentacle rising from its head, then a dazzling larval eel that looked like a transparent willow leaf, and a golden-tailed sea dragon with a bright cheek light that was much bigger than any he had ever seen in his nets.

Hollister recognized the throes of discovery in Beebe's voice when, at 1,500 feet, he was watching a jelly close enough to brush the glass and a large fish materialized as though from nowhere. It was suspended half in and half out of the beam, keeping its position with a slow waving of its fins. Beebe knew instantly that the creature was something wholly unknown, a discovery of the same magnitude as *Bathysphaera intacta,* but this time he was not fighting for his life in a wildly tumbling Bathysphere as he was in 1932 and he could make an indisputable identification. The pale, olive-drab-colored fish was two feet long, with no luminous organs that Beebe could see, small black eyes, and a large, underslung jaw. Its tail was a tiny knob, with vertical fins rising high above and beneath its body. After thirty seconds riveted to the window, Beebe grabbed Barton by the shoulder and pulled him over to confirm his sighting, which the younger man did by smiling and patting Beebe on the back. The fish was part of no known genus or species, so when Beebe proposed his description in the November–December issue of the *Bulletin of the New York Zoological Society,* he would name both, calling the creature *Bathyembryx istiophasma,* and also giving it the common name pallid sailfin. He placed it in a family called Cetomididae, of which only three species were known and twenty-four specimens captured, sixteen of them in Beebe's own nets. The ghostly fish hovered for over a minute, half in and half out of the beam, inspecting the Bathysphere and giving Beebe an exhilarating sense of success.

> The Sailfin was alive, quiet, watching our strange machine, apparently oblivious that the hinder half of its body was bathed in our strange luminosity. Preeminently, however, it typified the justification of the money, time, trouble and worry devoted to bringing the Bathysphere to its present efficiency. Amid nameless sparks, unexplained luminous explosions, abortive glimpses of strange organisms, there came, now and then, adequate opportunity to add a definite new fish or other creature to our knowledge of the deep sea.

The sailfin moved away like a fun-house apparition dissolving into the darkness, and Beebe gave the okay to continue the descent. At 1,900 feet, he thought he could still see the faintest tinge of gray light in the blackness outside his window, and speculated to Hollister that he had never been at that depth on so bright a day and that must account for the last whisper of the sun's presence. Barton held two fingers up in front of the window and Beebe could see their shapes backlit by the still-glowing sea. But then at two thousand feet there was no hint of the gray, nothing but the most absolute blackness, which meant every remnant of the visible spectrum was gone and Barton's fingers were invisible against the glass. Beebe flicked on the light at low power: its beam projecting into the colorless universe was a weird turquoise that deepened to a rich, velvety dark blue as it approached the edge of its range.

Sinking past two thousand feet with the searchlight turned off, the Bathysphere passed through clouds of pyrotechnics that became motes, shrimp, jellies, and fish when the beam again reached out to touch their once-inviolate world. At 2,300 feet, when Beebe was rhapsodizing about a large, unidentifiable form just beyond the light beam, Hollister interrupted him and insisted he listen to the whistle salutes from the world above that were acknowledging the new depth record. Beebe paused for a heartbeat and shot back, "Thanks ever so much, but take this: two very large *leptocephali* have just passed through the light, close together, vibrating swiftly along. Note, why should larval eels go in pairs?" And on down they went through 2,400 feet with the light off, to 2,500 feet, where Beebe told Hollister he would hold for a half hour. "How long?" she asked, incredulous. "A half hour," Beebe replied. Hollister then told him that their actual depth was 2,510 feet, and settled herself to listen.

Beebe ceded the window to Barton and switched the searchlight to full power, instantly warming the Bathysphere for the four minutes the Bolex chittered and whirred. The full 1,500 watts threw the beam out to what Beebe estimated was forty-five feet. It was startlingly bright, though of the same eerie white-turquoise shade as the light from the low-power lamp. The camera worked perfectly, and Barton withdrew from the window to reload while Beebe moved back into position to watch four odd fish swim in stiff, upright positions through the light. He was amazed by their brilliant scarlet jaws and heads, bright blue gills, and yellow bodies and tails, and wondered why an abyssal creature would have evolved such color, which could only be seen in the bioluminescence of a large predator or the light from an intruding Bathysphere. They were about four inches long, and might have been the reef fish called rainbow gars, but Beebe didn't make a positive identification. Keeping perfect formation and distance from one another, the abyssal quartet swam upward and out of the beam.

Still at 2,500 feet with the electric beam out, Beebe systematically studied the flashes of bioluminescence for a measured minute, counting up to forty-six over the phone to Hollister, and noting that most were pale yellow but a few were bluish. As he finished his inventory, he turned on the light and was overwhelmed with joy when he saw a pallid sailfin glide into the beam. It might have been the same one he saw a thousand feet above, or it might have been a different fish; but either way, it was real.

For a mere half hour, Beebe and Barton were denizens of the abyss, and their Bathysphere was a microscopic particle in an immense, alien universe. Through a glass window the size of a hand span, they lost themselves in a world of eternal darkness to which only they had ever been admitted. The steel that protected them from the freezing water and millions of pounds of pressure was uncomfortably cold, and the two men were constantly changing their positions to avoid its contact. They were reminded of their former lives only by Hollister's voice on the telephone, the hiss of the oxygen flowing from its tank, and the whine of their air blower, all of which represented the history and effort of thousands of men and women who invented the means to get them where they were.

When two hours had passed since they left the *Ready*'s deck, some instinctive alarm sounded in Beebe, who was suddenly tired and having difficulty keeping track of the flashes of light in the darkness. They were 140 feet short of a half mile, but he ordered the ascent. Later, Beebe would say the thought of going on to a full half mile never crossed his mind.

The sensation of rising hit them ten seconds after Beebe told Hollister to bring the Bathysphere up. A minute later, while they stopped at 2,470 feet for a rope-tie cut, he settled back into the cosmos beyond the window. He was immediately face to face with an anglerfish that darted into the beam to inspect the glass, hovering long enough for Beebe to claim another identification of a creature never before known to man. At first, he saw only its lights and their slight reflection on the skin of its back, but it swam into the beam, which was on at low voltage, and revealed itself. The fierce-looking little fish looked a lot like two other anglerfish Beebe knew from his nets, *Ceratias* and *Cryptosparas,* but not exactly. In his proposal of the description in the *Bulletin,* he would name it the three-starred anglerfish, *Bathyceratias trilychnus.*

As the three-starred anglerfish rejoined the unknown in the depths of the ocean, Beebe's euphoria of concentration faded and he slumped from the window. He and Barton performed another of their many grand shiftings of limbs, though the complaints of their muscles after more than two hours weren't eased at all and they knew they still had at least an hour to go before they would be free in the sunlight. Up through two thousand feet, Beebe continued his rambling inventory up the line to Hollister, but a flatness in his voice told her that he was suffering.

While stopped at 1,900 feet, Beebe detected the first faint evidence of sunlight brightening the smear of white lead at the window's edge. Just as he remarked about it, the Bathysphere began to move again and he snapped to Hollister, "Hold here." Beebe couldn't believe his luck: his weariness dissolved in the exhilaration of his description of a third fish new to science. He would call this discovery the five-lined constellation-fish, *Bathysidas pentagrammus,* when he fleshed out his description for the pages of the *Bulletin* two months later. In Beebe's memory, his few seconds with a living five-lined constellation-fish in its Stygian home would remain forever as among the loveliest of his life.

After the five-lined constellation-fish vanished into the darkness, Beebe and Barton endured another hour before reaching the surface. Both men were slightly euphoric from their extended time in the pure oxygen atmosphere, and Beebe's observations were cursory at best. They spent most of their time during the ascent shuffling around to ease cramps. Still, when they were finally back on deck after more than three hours in the Bathysphere, their arms and legs were numb and use-less. They needed help to crawl through the hatch and stand up. When the divers were steady on their feet, Beebe, Barton, Tee-Van, Hollister, Crane, Long, and the reporters followed Jimmy Sylvester to the bow, where he lowered one of his crewmen over the side in a bosun's chair with a bucket of whitewash and a brush. On the *Ready*'s weathered flank, the sailor painted the numerals 2 5 1 0, and John Long, mimicking a radio announcer's basso voice, said, "That, ladies and gentlemen, is a new world's record."

Four hours after the *Powerful* and the *Ready* reached the dock in St. George, Long wired the official National Geographic account of the descent to the society headquarters. The next day, the *New York Times* published it on the front page with 1932 file photos of Beebe, Barton, and the Bathysphere, and the article was given similar treatment by every major newspaper in the United States. The world's record led the story.

DR. BEEBE DESCENDS
2,510 FEET IN OCEAN

WASHINGTON, AUG. 11   The National Geographic Society tonight gave out the following account of Dr. William Beebe's descent to a new world's record depth of 2,510 feet below the surface of the ocean:

"Down where the blue begins, nearly half a mile below the surface of the Atlantic, Dr. Beebe and Otis Barton gazed upon eerie worlds of pres-sure and darkness never before seen by living man. In a steel bathysphere the intrepid adventurers broke their own world's record of 2,200 feet by diving to 2,500 feet off Nonesuch [*sic*] Island.

"Fellow scientists and representatives of the New York Zoological Society and the National Geographic Society, under whose auspices the exploration was made, listened from the deck of the converted barge *Ready* as Dr. Beebe enthusiastically reported by telephone scores of fish new to science and dictated thousands of words of description about little-known denizens of a world strange as Mars. Mr. Barton, with the aid of a special high-powered light, took motion-pictures of weird creatures that floated by the quartz eyes of the bathysphere.

". . . 'I have never seen so much stuff in my life and new stuff, too,' Dr. Beebe said upon emerging from the bathysphere. 'Much of it is entirely different from that which we observed during previous dives. It is the silliest thing in the world to attempt to describe it in a few words, but we saw more fish and larger fish than during any other dive.

" 'Every dive convinces me of the futility of trying to get the true idea of deep-sea life through dragging nets. Many deep-sea creatures are such rapid swimmers that they can easily get away from nets. One of the most amazing finds of the day was a flesh colored fish which I observed near the 2,500-foot level. We have never caught any fish from this depth. We also observed schools of rare lampanyctus, hatchet fish and thousands of tiny squids.'

"Asked why he came up when he did, Dr. Beebe replied that the exact record was not as important to him as observations he wished to make at different levels. The bathysphere and its air-conditioning equipment were functioning perfectly, but it was decided to come up and make a deeper dive later, perhaps early next week, at which time further observations will be made to check the data obtained today."

Eighteen

HALF MILE DOWN

The only other place comparable to these marvelous nether regions,
must surely be naked space itself, out far beyond atmosphere,
between the stars, where sunlight has no grip upon the dust
and rubbish of planetary air.

William Beebe

The day after the descent to 2,510 feet, a Sunday, Else Bostelmann sat at her drawing board in a sunny alcove of the New Nonsuch laboratory overlooking Castle Harbour and drew while Beebe told her what he had seen in the abyss. She worked with charcoal for the sketches; later, as she had with dozens of other specimens from Beebe's nets and descriptions from his descents, she would render the creatures in oils and watercolors. They spent most of the day on the billowing fins and tail bulb of the pallid sailfin, the shape and appendages of the three-starred anglerfish, and the array of lights on the oval body of the five-lined constellation-fish, working from the notes Beebe had given Hollister and his memory of the sightings. Bostelmann captured the strange upright formation of the scarlet, blue, and yellow fish Beebe was calling rainbow gars, and experimented with her palette until she captured the odd colors to match those that flashed through the searchlight beam at 2,500 feet.

Beebe and Barton were exhausted after their three-hour confinement in the Bathysphere, so for two days they relaxed, even though the weather was perfect for diving. For Beebe, rest meant field trips, and Monday morning he led a helmet-diving excursion to the northern reefs with Tee-Van, Hollister, Crane, Long, Ramsey, and Bass. Back in the Kingdom of the Helmet, Beebe explored the submerged cap of Bermuda's mother volcano at forty feet, a benign wonderland so different from the darkness of the abyss that it was hospitable by comparison. The living colors of sponges, urchins, corals, and fish, and the aquamarine shimmer of the water itself, transported him back to the day when he had made his first helmet dive nine years before. Beebe had been seduced by the ocean, but he had no regrets about his decision to throw himself wholeheartedly into its embrace. Since the moment of his rebirth in a remote anchorage among the bare, windswept humps of the Galápagos, his obsession with the sea had not betrayed him, but had ripened instead into exactly the kind of joy he believed that a benevolent universe granted to men who

persevere with courage. Beebe knew that the world was indifferent to him, but thought that by making choices and living with their consequences, he became part of the perfect rhythms and patterns he had first seen so long before as a child enchanted with nature.

The helmet dive was a roaring success, with sightings of sharks, moray eels, and barracuda, capped by a champagne ride back to St. George in the *Skink*. The next morning, Tuesday, Beebe took the *Gladisfen* for a test run after her overhaul and made a dozen passes through the same patch of the Atlantic as that of the deep dive, towing nets at depths of 300, 600, 1,200, 1,800, 2,400, and 3,000 feet. He hoped against hope that one of his newly discovered treasures would surrender itself, but there were no pallid sailfin, three-starred anglerfish, or five-lined constellation-fish in the trawls. Still, everyone was happy with the prizes of the day: a pair of ten-inch scimitar-mouths, *Gonostoma elongatum;* and a black swallower, *Chiasmodon niger,* a bizarre creature that can eat a fish larger than itself by distending its body. Though every haul brought something marvelous from the icy depths, Beebe was amazed at how little he had been able to capture in nets compared with the riot of life he knew was really down there. He couldn't wait to get back.

While Beebe was busy observing, capturing, and naming, Barton hunkered down at the St. George Hotel with Dority and wondered whether he should go back to New York. The Abercrombie & Fitch photo lab in Hamilton had developed his plates and movie film from the 2,510-foot descent and he had come up with absolutely nothing, despite the high-powered searchlight and some experiments he had tried with a recently invented flashbulb. Beebe's spectacular discoveries of deep-sea creatures were good for Beebe, and the new world's record was making all the papers. Barton, on the other hand, was still only mentioned as an assistant and never in the headlines, and he wasn't looking forward to another deep dive without any film to show for it. After five years with Beebe, Barton thought he knew the man pretty well and he was sure that nothing could keep the naturalist from trying for a half mile or more on the next descent.

Though he had already done it dozens of times, Barton spent a few minutes computing the force of the water and the weight of the cable and Bathysphere as it went deeper into the sea. At 2,500 feet, the pressure demon would exert 1,129 pounds of force on every square inch of glass and steel, for a total of 11,644,000 pounds on the whole sphere; at 3,000 feet, 1,360 pounds per square inch, or 14,026,429 pounds. Barton remembered his final test with the new windows at the Watson-Stillman shop back in May when he had run the pressure up to two thousand pounds per square inch. Although the windows had held, they would be near that limit during a dive to the end of the cable. Who knew what the previous deep dive had done to the integrity of the quartz glass and steel that stood between him and eternity? Barton was certain that the cable was strong enough, but he was worried about the winch drum, which had cracked in 1932. At a half mile, the six-and-a-half-ton weight of the Bathysphere and its cables would be near the extreme limit from which the *Arcturus* winch and the *Ready's* two boilers could haul them back from the abyss. If the winch failed to lift the load, Sylvester could radio for another tug and winch from the navy yard at the other end of Bermuda, but there would be little chance that help would arrive before the oxygen tanks ran dry.

Late Tuesday afternoon, when Barton's fertile imagination was wearing him out and he could stand no more thoughts of a useless death in the darkness of the abyss, he left the hotel alone and walked down the hill through the pastel facades of the old town to the Darrell and Meyer wharf, where the *Ready* lay easy on her lines like a giant, napping animal. There was no sign of Sylvester or anyone else on the barge, so he boarded and went to the foredeck, where the Bathysphere sat baking in the sun. The hatch cover was off, and Barton removed the rope web woven from bolt to bolt to cover the opening and crawled inside. On a calm summer day at dockside, without the adrenaline of a dive blotting out his senses, the aromas of sea salt and sweat wrapped around him like a second skin. The Bathysphere was an oven under the high, hot sun. Barton thought of the chains of fate and circumstance that had brought him to that point, remembering his timid approach to Beebe at the zoo and wondering if he should have gone it alone after all. Barton had been much more comfortable with the *idea* of William Beebe when the great naturalist was just another hero in his pantheon of adventurers than he was with the real Beebe—a paternal figure who seemed barely able to tolerate him. Beebe had not invited him to go along on the helmet dive or the trip offshore on the *Gladisfen,* but that was nothing new. Barton had built and paid for the Bathysphere, screwed his courage to the sticking point on eighteen frightening descents, and taken the first historic photographs of the abyss, but after all that, Beebe still didn't like him. He was never part of the Crowd.

The walls of the Bathysphere were stripped of the oxygen tanks, chemical racks, searchlight, and the other diving equipment, so the interior felt spacious to Barton as he examined the arcing steel walls that had been the key to his vision so many years before. He reached out and rapped his knuckles on the center window, remembering John Butler's excitement when he had found E. E. Free and quartz glass, and looked over his shoulder at the big brass nut of the stuffing box that had solved another of the puzzles of descending into the deep ocean. What the hell, he thought. No one else in the world had advanced as far as they had into the technology of deep descents, so a record of three thousand feet would stand for a long time, bringing with it a permanent place in history no matter what else life handed him from then on out. Barton owned part of the record of 2,510 feet, and if Beebe was going to the end of the cable, he would be a fool not to go with him.

At 7:25 on the morning of August 15, the *Gladisfen* and the *Ready* slipped out of St. George Harbour and into the flat, calm Atlantic, where the sun on the eastern horizon flashed off the highlights of the tug's white wheelhouse, the wire rigging, and the light blue globe on the bow of the barge. The procession turned right at the mouth of the channel, passed St. David's light, Nonsuch Island, Gurnet Rock, and the rocky guardians of Castle Harbour, and made for open water at a stately three knots. Except for the feather touch of the warm air at that speed, there wasn't a hint of a breeze, no swell at all offshore. Aboard the *Ready,* Beebe smiled at his luck.

Two hours later, Jimmy Sylvester on the *Gladisfen*'s bridge rang down slow ahead to his engineer. The barge's weight on the end of the tow rope quickly slowed the tug to a crawl that held them steady against the southeast current. As near as Sylvester could determine from triangulation with the lighthouses on shore and a plot on his chart, they were at the same spot on the surface from which Beebe and Barton had made their descent four days before. Sylvester turned the helm over to his mate and went aft to board the *Skink,* which had followed them out with Arthur Tucker at the wheel, and transferred to the *Ready* to command the deck crew. As he approached the barge, he saw the celebratory numerals 2 5 1 0 still on the bow and he yelled at one of his men to wash them off for luck.

The burbling sound of the tug's engine dropping to low rpm had galvanized the well-drilled crew on the *Ready,* and preparations for launching the Bathysphere were completed by the time Sylvester climbed aboard. Beebe and Barton were already inside, Tee-Van was hammering home the bolts, Crane was standing by the winch ready to tally the depth, and John Long, oddly quiet with no reporters to supervise that day, watched from the bow. The cable was festooned with the banner of the National Geographic Society; above it, for the first time on any Bathysphere dive, waved the American flag. Hollister, at her station on the starboard rail, made a note in the log, "10:02 Wing bolt going on," and through the open phone circuit she heard Beebe say, "Well, Otis. Here we are."

For the next hour, as the Bathysphere made its downward plunge out of the world of the sun, Beebe watched Else Bostelmann's paintings come to life again before his eyes. Aurelia, salps, siphonophores, and pteropods, copepods, and other plankton dominated the upper layers down to eight hundred feet, the depth of their first dive four years before. Thank goodness, Beebe thought, we were able to go deeper. The dark blue belt between the disappearance of other colors and the pure black of the abyss reclaimed the Bathysphere as it descended through one thousand feet, where illuminated creatures began to flash and sparkle in the darkening sea. Beebe saw something with six lights in a row, then a large, pale green nebula, then a fish right at the window with its fangs and nostrils lit from within like a perfect Halloween mask, and another with the luminous toothy smile and eye holes of *Alice in Wonderland*'s Cheshire Cat. Old friends appeared in the beam of the searchlight as though on cue: a pair of silver hatchetfish, *Argyropelecus,* with their bulging eyes and tinseled flanks; a swarm of pale arrow worms, *Sagitta,* devouring a small fish; and a shining green sea dragon, *Lamprotoxus.* At 1,600 feet, Beebe solved the puzzle of odd bursts of brilliance that he had seen so many times before and couldn't explain. A large shrimp drifted in front of the window, turned sideways and exploded, and he realized that he had witnessed the defensive light cloud of a deep-sea shrimp. The flash was so bright that Barton saw it illuminate Beebe's face and the inner sill of the window.

The white lead sealer around his window caught the last wink of sunlight at 1,900 feet, and the Bathysphere continued on into the eternal darkness. Gazing into the bizarre, violent, crowded world of the abyss, Beebe was in heaven again. Barton patiently looked over his shoulder and kept an eye on the oxygen gauge, humidity indicator, and stuffing box, and relayed the readings to Hollister for the log. At two thousand feet, the conditions were perfect: oxygen at 1,650 pounds of pressure in the first tank was seeping out at one liter per minute; humidity held at a comfortable 60 percent; and the temperature stood at 86 degrees, even though the steel was cold to the touch. Every ten minutes or so, Beebe gave Barton and his camera the window. Though the film from the unmanned test dive with the timer camera had produced recognizable images, Barton had come to the conclusion that photography in the abyss was largely a matter of luck, and he wasn't expecting much. Still, he tried to improve his chances with a new wide-angle lens to keep everything in focus through a band four feet wide in front of the window, knowing that everything he learned, whether from success or failure, would be important to explorers who followed him.

Passing through 2,500 feet, Beebe switched the light on at low power just as the outline of something at least twenty feet long ghosted through the dimness at the end of the beam. Seconds later, the featureless shadow returned for another split-second passage through the most distant rays of the light. Beebe couldn't possibly make an identification, but he knew he had seen something big and speculated that it was a whale or a whale shark. He pulled Barton to the window, but the creature was gone. They remained head to head looking together into the abyss. Seconds after the sea monster had vanished, Barton saw the first *Stylophthalmus* ever seen alive. To his chagrin, Beebe somehow missed the strange little creature with eyes on the ends of periscope stalks almost a third as long as the entire body. From Barton's sight description, Beebe knew that it was the larval form of the golden-tailed serpent dragon, *Idiacanthus,* which he had been studying from net specimens for five years.

At 2,600 feet, Beebe held the ear pad of the phone set away from his head to share the sounds of the whistles and horns above celebrating another new record, but seconds later, at 11:06:21, he completely ignored hitting the mark that he had been promising the world since his first season of diving, 2,640 feet. A half mile. The next thing Beebe said, according to Hollister's entry at 2,650 feet, was "Millions of sparks hit the window."

Beebe had not discussed his intentions with Barton, who had assumed correctly that the destination for their descent was not a half mile but the end of the cable. As they stopped at 2,700 feet for a rope tie, Barton pointed his flashlight beam up at the stuffing box and told Beebe that only a half inch of the Okonite cable had been forced through, which was good news. Oxygen flow was holding steady at one liter per minute, with only half of the first tank gone, and the humidity hadn't changed since the last instrument scan, which meant the chemicals were keeping up with the carbon dioxide and moisture. "All okay," Barton said, his voice a whisper barely audible above the sound of the blower. Beebe never broke his concentration at the window, and Barton leaned over his shoulder and watched the once-forbidden world of the abyss slide by in the beam of the searchlight. Beebe's voice carried no hint of fatigue as it flowed up the line to Hollister.

So black outside, and what lights! A fish with a long, slender, pointed tail. This is a big fish. Beam on. Here's a telescope-eyed fish, it's Argyropelecus, and its eyes are very distinct. Barton sees something like a huge necklace of silvery lights. Now another big shrimp. Beam off. Marvelous outside lights. Water filled with lights, more so than on our last dive to 2,500 feet.

Beebe took a breath. Hollister told him that the Bathysphere was passing through 2,900 feet, and he went on.

Now a curved, pale-green fish, light under eye, eye lighted up by it. It is crescent-shaped. The fish at least 3 feet long. 5 inch Myctophids, swimming so slowly that I can see whole light pattern. Several close lines of lateral lights, and constantly lighted plates. *Lampadena,* maybe but not sure, try to look up species. Suddenly, not a flash in sight. Now a Siphonophore, a big one. Oxygen 1,400 pounds, barometer 76, temperature 77, humidity 62 percent.

"Director," Hollister blurted. "You are now at three thousand feet."
As Beebe and Barton hung in the darkness below, Jimmie Sylvester watched the last turns of 3,200 feet of cable unwinding from the winch, exposing the wooden core of the drum, and he told Hollister to tell Beebe he could go no deeper. Beebe replied that he wanted a half hour to observe, but when she relayed the message to Sylvester, the captain said absolutely not. A slight swell had built on the midday breeze, the *Ready* had already picked up a waltz-time roll, and Sylvester wasn't at all sure what would happen when the full weight of the Bathysphere and

all its cable was submitted to the force of the wallowing barge. But he didn't want to wait to find out. "Five minutes," Sylvester said. "Then they come up." Hollister relayed the message to Beebe, who grudgingly agreed and returned to his narration of the fantastic story beyond the window: "Long lace-like things again, salplike with big head and long slender tendrils. Now another one."

Just before Sylvester's five minutes had elapsed, Beebe rubbed his cold hands together to get the circulation going. Then, in a private ceremony, he held his penlight in his mouth and signed his autograph three times on a piece of notebook paper.

WILLIAM BEEBE

3,028 FEET   AUGUST 15, 1934

Then the Bathysphere dropped slightly as the winch dogs were released above and Beebe and Barton felt the familiar heaviness that signaled the beginning of the trip back to the surface. Before they had risen a hundred feet, though, Beebe was describing a strange bright flash outside when a sickening metallic report rang through the phone line and the Bathysphere shuddered and lurched to a halt. Beebe stopped talking, Barton held his breath, and they both prepared for the worst.

Though they were never friends and their tolerance for each other had worn thinner with familiarity, Beebe and Barton had known all along that they would die together if a window shattered or the hatch seal failed or the cable broke. They also knew that dead or alive, they would be remembered together forever as the first humans to descend into the depths of the ocean, no matter whose name was in the newspaper headlines.

Beebe and Barton exhaled as Hollister's voice again flowed down the line to tell them that a rope the crew was using to guide the cable onto the winch had snapped and caused the frightening sound. Everything was fine. Captain Sylvester was continuing the ascent. Beebe turned back to his observations and saw a spectacular galaxy of sparks in the middle distance. His first words up the line after the heart-stopping moment of terror were: "A lovely light."

An hour and twenty minutes later the Bathysphere broke the surface, rose dripping in the air with the sunlight flashing off its windows, and settled safely onto the *Ready*'s deck. Tee-Van and Long pulled the wing bolt, then the hatch cover, and Beebe and Barton painfully crawled out into their natural world. They had been inside for three hours and three minutes, and had to be helped to their feet. When they were able to stand alone, and the whistles and handshakes ended, Sylvester lowered a sailor over the side to paint 3 0 2 8 on the bow and John Knudsen took the commemorative photograph of the Bathysphere and its crew. In it, the flags of the United States and the National Geographic Society snap in the fresh breeze on the cable, and the hatch cover, a seat cushion, and a visor are seen lying on the deck. Beebe is standing on the right, bareheaded, with his outstretched arm on the Bathysphere, and Barton is in the center of the frame looking directly at the camera from under his lucky cap with his hands behind his back. Both men are squinting against the brightness of the sun and their expressions are identical masks of sublime exhaustion, carrying a combination of weariness, delight, and relief.

Epilogue

TITANS OF THE DEEP

*Beebe seems more assured in his status as a pioneer of ocean ecology—
largely because of his systematic sampling, his many indisputable
specimens, and his ability to inspire others to follow him into the abyss.*

Robert D. Ballard, deep-ocean explorer, in *The Eternal Darkness:
A Personal History of Deep-Sea Exploration*

*In 1985, when entire life histories may be photographed in the abysmal
depths, these first blotches of light on a black field may be treasured
as the work of a pioneer who opened up an environment that is as
black and remote as interstellar space.*

William Gregory, ichthyologist, on the footage shot
by Otis Barton during an unmanned dive to three thousand feet

William Beebe never again descended beyond sunlight. Eight days after their dive to the end of the cable, he and Barton reached 1,507 feet and spent more than two hours observing and photographing, but neither man wanted to risk going deeper again. Beebe was confident enough in the safety of his craft in shallower depths, though, that he let the other members of his expedition experience the wonders of the dark world beneath the sea. On the afternoon of August 15, Hollister and Barton went to 1,208 feet, setting a new record for a woman diver that would stand for three decades; on August 27, Beebe took Tee-Van with him to 1,533 feet; on September 11, Beebe and Crane went to 1,150 feet; then Beebe's young assistants, Perkins Bass and William Ramsey, descended to 550 feet alone. And finally that day, Beebe and Barton together made the last dive of the Bathysphere to 1,503 feet. Three hours later, William Beebe and Otis Barton said good-bye at the Darrell and Meyer wharf and never spoke to each other again.

The two men continued to appear together in newspaper stories and newsreels, but their estrangement grew more bitter and eventually broke through the veils of privacy and propriety. Barton was promoting himself as a filmmaker and wanted to remind his potential audiences that he had been to the abyss, seen real sea monsters, and was alive to tell about it. When he was left out of stories in newspapers and magazines, he wrote letters to the editor to correct the record and make sure he wouldn't be forgotten the next time.

Beebe was outraged by Barton's letters, which contained errors and exaggerations about what he had seen in the abyss, primarily because they cast doubt on Beebe's own observations and fanned the brush fires among the already skeptical tribe of ichthyologists. Unfortunately, Beebe had made no further sightings of the untouchable Bathysphere fish, the pallid sailfin, the three-starred anglerfish, or the five-lined constellation-fish, and none had come up in the trawls of any other expeditions. Allowing Barton to comment in any way on the scientific discoveries of the Bathysphere expeditions was nothing but trouble, and Beebe and Tee-Van pulled out all the stops to do damage control. "Barton, we all realize now is not a villain but a dangerous fool," Beebe wrote to John Long in late September 1934. ". . . by the grace of Providence we have just killed a letter of his to the *Times* in which he misidentified all fish . . . which would have made a bad mess of things."

Beebe made good on his deal with Gilbert Grosvenor and two weeks after he left Bermuda, he turned in the promised article for *National Geographic,* "A Half Mile Down: Strange Creatures, Beautiful and Grotesque as Figments of Fancy, Reveal Themselves at Windows of the Bathysphere," with twenty-nine photographs and drawings and sixteen of Else Bostelmann's paintings. It ran in the December 1934 issue of the magazine. Grosvenor was delighted, and he sent Beebe his copies with a letter of praise:

> It seems to me that this is the finest contribution you have yet written for *The Geographic.* The pictures which Mrs. Bostelmann made under your direction are superb, very dramatic and very informative.
>
> As I was in Europe at the time of your deep-sea dives last August, I had not the opportunity then to congratulate you on the superb records you achieved. I hope the officers and members of the New York Zoological Society are as much gratified by what you accomplished during last season's work as are the officers and members of the National Geographic Society.

The text of the *Geographic* article also became the climactic chapter in Beebe's book, *Half Mile Down,* which appeared in bookshops in time for Christmas that same year. *Half Mile Down* was an immediate best seller, which was a good thing for Beebe, who was down to his financial reserves, as the Depression was at its nadir and he still owed money for expedition expenses. Beebe wrote to John Long to ask him to approach Grosvenor for help. "Please consider this private, dear Long. The expedition has run $2,700 over budget. Mrs. Beebe and I can stand it, but it will use up about all our money. Is there any chance that G.G. can come up with half?" LaGorce, speaking for Grosvenor, turned him down and advised him to get the money from Barton. Beebe replied that he was "afraid to accept as I do not wish to be under any obligation to him in any way."

Beebe's light shone so brightly that winter that Barton eventually gave up trying to be the leading man in the Bathysphere drama and retreated to Panama to work on his own movie. With a new sidekick, a South African wool merchant turned screenwriter named Ned Ewer, Barton chartered a sailboat with a crew of three and went looking for sharks, sawfish, and giant manta rays in the Caribbean and on the Pacific side of the isthmus. Barton disappeared from New York for almost three years, skipping the opening of another Bathysphere exhibit at the American Museum. When he was in the United States, he began living alternately in New York and Long Beach, California. He eventually finished his underwater adventure movie, which was released by Grand National Pictures at premieres in Chicago and New York and played at theaters around the country.

According to the lobby posters, *Titans of the Deep* starred Dr. William Beebe and Otis Barton, with John Tee-Van, Gloria Hollister, and Jocelyn Crane. The movie was a patchwork of surface shots of Bathysphere launches, staged views through the windows, footage of shark and barracuda attacks, and other underwater scenes with only a hint of a plot. A *Film Daily* reviewer, probably a flack, said it was

> a really remarkable underseas picture showing monsters of the deep at close range and built up into an exciting and suspenseful piece of entertainment without any hoke. Other entertainment features are a girl going down with a diving helmet and a rifle, with which she shoots an immense barracuda. And she is a darned attractive gal, too. One of the men also goes down in a diving suit equipped with a camera, which he loses as a great shark swallows it and he kills the shark with a spear. A fight between a giant lobster and an octopus on the ocean bed is a real thrill.

Not all the reviews were so kind. The movie was a bust, and *Titans of the Deep* closed three weeks after it opened.

Beebe finally saw *Titans of the Deep* at a matinee during the last week of its run in New York. That night he wrote letters to the *New York Times,* the *New York Sun, Science* magazine, and a half dozen other publications with his final public pronouncement on Otis Barton.

> I would like to correct a rather vital misconception in regard to the film now running in New York City and elsewhere in the country, called

"Titans of the Deep." This is being credited to me and my associates, whereas neither I nor any member of my staff of the Department of Tropical Research, nor any one connected with the New York Zoological Society, had anything to do with it.

At the very beginning are shown a few authentic shots of the Bathysphere, but all the rest of the film is the work of Mr. Otis Barton, and was taken in Panama at his own expense and with no relation to the Bathysphere. I never saw any of it until it appeared on a New York screen. Together with my staff, I would like completely to dissociate myself from this motion picture and to have it known altogether as the work of Mr. Barton.

BEEBE

William Beebe was fifty-seven years old when he made his last dive in the Bathysphere. After *Half Mile Down,* he wrote four more books, two about his oceanographic expeditions off California in 1937 and 1938, and one about his Venezuelan adventures. In his last book, *Unseen Life of New York: As a Naturalist Sees It,* published in 1953, Beebe ended his career as he had begun it, by describing nature in his own neighborhood. The Department of Tropical Research survived the Depression, but after World War II Beebe left the sea forever and went back to the jungle. The publicity surrounding the Bathysphere dives had elevated him to the heights of international celebrity, but he was still a sitting duck for accusations of fraud and shoddy science, and he was tired of it. Though he published dozens of seminal papers on abyssal creatures from his nets and observations, his antagonists never really let up on his sight descriptions of *Bathysphaera intacta, Bathyembryx istiophasma, Bathyceratias trilychnus,* and *Bathysidus pentagrammus.* A reviewer of *Half Mile Down* wrote in the ichthyology journal *Copeia:* "I am forced to suggest that what the author saw might have been a phosphorescent coelenterate whose lights were beautified by halation in passing through a misty film breathed onto the quartz window by Mr. Beebe's eagerly appressed face." Unfortunately for Beebe, none of the new fish he discovered from the Bathysphere has ever been seen, dead or alive, again—although any oceanographer knows that doesn't mean they don't exist.

In 1949, Beebe set up a permanent field station in Trinidad, which he named Simla, after the town just below the great hills in northwest India where Kipling's Kim is initiated into the mysteries of the Great Game. He was still married to Elswyth Thane, whose own literary career flowered into thirty-three books and two plays. But though they maintained a farm in Vermont and an apartment in the Hotel des Artistes in New York, Dr. and Mrs. Beebe spent very little time together. During most of the last decade of Beebe's life, Simla was his home and Jocelyn Crane was his most cherished colleague and constant companion. The Bermuda expeditions had launched Crane into a promising career as an ichthyologist, and she collaborated with Beebe on several major publications, including a landmark monograph on dragonfish. Taking a lesson from Beebe, though, Crane followed her passion rather than observing academic boundaries, and she abandoned fish to study marine invertebrates. She published a steady stream of respected articles, and her massive book on fiddler crabs remains the definitive work in the field. When Beebe became director emeritus of the Department of Tropical Research on his seventy-fifth birthday in 1952, Crane succeeded him as director.

William Beebe died on June 4, 1962, and was buried quietly in a jungle cemetery in Port-of-Spain, Trinidad. As news of his death made its way to Washington and New York, obituaries and eulogies flew from typewriters around America and Europe, including these words from Fairfield Osborn, president of the Zoological Society and the son of the man who gave a fervent young naturalist a job in 1899.

> He dared to write in his own fashion at a time when it was thought that all scientific writing must be purely methodical. Time has changed all this and today many a scientist and naturalist endeavors to write more colorfully and interestingly. None can excel the earlier adventurer, Beebe, who dared to live up to his vision that the wonders of the natural world belonged to everybody.
>
> Only those who worked intimately with him realized how intensely he believed in accuracy and in the provability of any observation. In this regard he was a true scientist, superbly equipped with powers of observation which frequently were an amazement to his associates. His goal was to gain fuller understanding of the living environment as a whole. His work contributed in many respects to a better understanding of evo-

lutionary processes but above all it opened new pathways to the science of ecology. There is no "end" to accomplishments such as these—their influence will run on through time.

Beebe's legacy as a pioneering ecologist rivals that of his contributions to ornithology, marine biology, and oceanography. The notion that an organism can only be understood when its surroundings and neighbors are taken into account was so radical during his time as to have been completely absent from most scientific discussion. The term "ecology" was not in common use when, in 1909, Beebe published a twenty-page article in *Zoologica* entitled "The Ecology of the Hoatzin." His field research on this large, crested, South American bird showed how local conditions played a major role in its evolutionary success.

Ecology is interdisciplinary by definition, and Beebe's innate curiosity unhampered by academic boundaries inspired him to roam freely as an observer of the natural world. He instinctively noticed the bridges of dependency and relationships among animals, and tailored his observational strategies to take them into account. His books and papers are peppered with references to the joy and value of scrutinizing twenty square feet of snow, a single tree in the forest, a thousand acres of jungle, or a cylinder of the Atlantic Ocean eight miles in diameter. Beebe's ecological studies were not systematic by modern standards, but he was on the right track. Legions of ecologists ever since have acknowledged him as one of their most important scientific ancestors. Rachel Carson, one of the first naturalist-authors to clarify ecology to general audiences, dedicated her landmark book, *The Sea Around Us,* to Beebe in 1951. "My absorption in the mystery and meaning of the sea," she wrote, "have been stimulated and the writing of this book aided by the friendship and encouragement of William Beebe."

BEEBE'S CROWD

Everyone from the Bermuda expeditions remained on the staff of the Department of Tropical Research, doing what they called "rainy day" work on a shoestring budget, until the war brought the curtain down on field science for five years. When that darkness lifted, everyone came ashore.

In 1941, Gloria Hollister married Anthony Anable, a metallurgical engineer from New York. Within a week of Pearl Harbor she joined the American Red Cross. In 1953 she cofounded the Nature Conservancy, and as its chairperson led a campaign to purchase and preserve its first wilderness, the Mianus River Gorge in Connecticut.

John Tee-Van became the executive secretary of the New York Zoological Society in 1942 and its general director in 1956. When William Bridges published his history of the society in 1974, the cover bore a photograph of Tee-Van holding a giant panda in his arms. Else Bostelmann enjoyed a long career as an artist at *National Geographic Magazine* and an illustrator of children's books; her paintings of abyssal creatures from the Bathysphere expeditions still find their way into exhibits at aquariums around the world. After Beebe died, Jocelyn Crane married a Harvard zoology professor, Donald Griffin, and lived in New York and Boston. When she died in 1998, her husband scattered her ashes over the Atlantic, just south of Nonsuch Island.

BARTON

Otis Barton never really left the ocean. After *Titans of the Deep,* he worked on another underwater movie in Panama and felt out life in Hollywood until the war pulled him into its maelstrom. He served as a navy photographer in the Pacific. Barton's trust fund had weathered the Depression, and when the war ended he went back to New York with the idea of resurrecting the Bathysphere to relive, without Beebe, his triumphant years off Bermuda. The Zoological Society was willing to let him use the Bathysphere, but told Barton that they had loaned it to the navy, which used it to measure the effects of explosives underwater, so it probably wasn't fit for deep diving any more.

When that plan fell through, Barton designed an entirely new craft that he called a Benthoscope, which was an improved Bathysphere built especially for photography, with walls two inches thick to allow a one-mile descent, an inside diameter of five feet, two windows, and two powerful searchlights in external housings. The Benthoscope would still be tethered to the surface by a cable, but the craft could also be rolled along the seafloor on a cradle with two front wheels, each six and a half feet in diameter, a smaller roller in the rear, and a sea anchor as a steadying drogue. In the fall of 1948, Barton went back to Watson-Stillman, the builder of the original Bathysphere, and agreed to pay $16,000 for the new casting, windows, and fittings of his Benthoscope, to be delivered in the summer of the following year. He would build the cradle himself.

Barton was planning to ship the Benthoscope to the Bahamas to dive in familiar waters when he met Maurice Nelles, a biologist from the University of Southern California. Nelles, who was also an engineer interested in undersea vehicles, said that the Allan Hancock Foundation would be delighted to have Barton and his Benthoscope join their research team. The foundation had already sponsored major expeditions and had put together a superb collection of marine specimens, and Barton was thrilled to be included. He told Nelles that he wanted to make the first test dives alone, but if the Benthoscope proved to be safe, he would let foundation scientists use it themselves.

In the spring of 1949, Barton bought a car and drove to California. At the Hancock Foundation workshops in Santa Maria, he spent three months building his cradle and fitting the electrical cables to his Benthoscope. In late July, Barton trucked the finished craft to Long Beach and loaded it on a chartered salvage tug for test dives. Barton and Nelles sailed aboard the Hancock Foundation's yacht, *Valero IV,* with Captain and Mrs. Allan Hancock, and enjoyed the luxuries of radio, television, modern depth-sounding equipment, and a fully equipped marine laboratory. Barton's flotilla also included the 105-foot schooner *Monsoon,* owned by Jack West, an advertising man who was thrilled to be adventuring with one of the heroes of the abyss. West was just tagging along, but he also had connections to the Los Angeles media and managed the publicity for the expedition. *Monsoon* carried West, his wife Carolyn, their three children, and a half dozen newspaper reporters from Los Angeles.

In early August, Barton's fleet set sail to try out his crawling cradle in shallow waters off Santa Barbara. On the first unmanned test dive, the roll of the barge in a light swell sent a series of jerks down the cable that sheared the bolts holding the cradle to the Benthoscope. Barton's wheels are still there on the bottom. Undaunted, Barton sailed north to the deeper water of the Santa Cruz basin, and on August 12 he sent the Benthoscope down to six thousand feet with an automatic camera aboard. The sphere came up dry, but the film was blank.

The next morning, almost fifteen years to the day after his last dive in the Bathysphere, Otis Barton again descended into the abyss, this time alone. The lights shorted out, the electric fan over the chemical trays failed, and the phone line went dead, but he made it down to 2,300 feet before he called a halt to the dive. Barton and Nelles tried to sort out the electrical problems, but failed. On August 16, Barton decided to dive without lights. He spent a terrifying hour in total darkness, but reached 4,500 feet, breaking the record of 3,028 that he shared with Beebe. Later that summer, Barton made two contour dives in shallow water with De Witt Meridith, a Hancock Foundation biologist, and called it quits for the season.

Barton put the Benthoscope into storage in Long Beach. For the next two years, he turned his attention to his next big idea: a machine to climb trees from which he would photograph animals that live in jungle canopies. He built an aluminum cage big enough to hold a man, rigged it with two hand winches, cables, and brakes, and in the spring of 1951 set off for French Equatorial Africa. The tree climber actually worked, but it was too slow and so noisy it scared off the wildlife, so Barton left Africa with a new big idea: why not photograph jungle treetops from a balloon?

Barton forgot about his jungle balloon a few days after he got back to New York, though, when he got a call from an editor at *Life* magazine offering him $10,000 to make another Benthoscope dive. In return, *Life* wanted exclusive rights to the story and photographs of the expedition and the creatures of the abyss. Barton accepted the offer without hesitation, rushed to California, rounded up Maurice Nelles, Jack West, and the *Monsoon,* and dusted off the Benthoscope, which was in pretty good condition after two years in a warehouse.

For two weeks in late September 1952, *Life* photographer Jay Eyerman and writer Sandy Thomson patiently waited while everything that could go wrong did. Valves, lights, telephones, cameras, and cables failed or broke during tests, and in two weeks Barton managed to make only three dives. The first, to two thousand feet, ended when oxygen free-flowing into the sphere forced an emergency ascent. The second, to 1,400 feet, was aborted when the lights went out. And at 3,331 feet on the third dive, Barton yelled, "Oh, Dr. Nelles, Dr. Nelles," and the phone went dead. Nelles ordered an emergency ascent, but as the sphere came up, the electrical wire became so tangled in the lifting cable that it had to be cut off to keep the winch from jamming. It took forty minutes to get Barton up, during which time everyone was sure that he was dead. Finally, the Benthoscope was hoisted on deck; through the portholes the crew could see that at least it was not full of water and also saw a faint glow that looked like a flashlight. They hammered off the hatch bolts and Barton emerged, yawning. The banging had woken him up, he told them. When the lights went out, he figured he might as well curl up in a blanket and go to sleep.

That was Barton's last descent, and the Benthoscope was retired. The story and photographs of the Benthoscope dives ran in *Life* in October 1952. Until late the next year, when Auguste Piccard and his son Jacques descended beyond ten thousand feet in their untethered Bathyscaphe, the *Trieste,* Otis Barton was the king of the abyss. The Piccards ran out of money and eventually sold the *Trieste* to the U.S. Navy. In January 1960, Jacques Piccard and an American submarine officer, Don Walsh, descended in the rebuilt *Trieste II* seven miles to the bottom of the Challenger Deep, a section of the Marianas Trench off Guam named in honor of the ship aboard which the modern science of oceanography was invented.

Among deep-sea explorers, the Bathysphere and Benthoscope dives were the founding legends, and Barton became a kind of eccentric grand old man of oceanography. His money held out, and for almost sixty years he roamed from one project to the next, collecting specimens for the California Academy of Sciences, attending the fiftieth anniversary commemoration of the Bathysphere descents in Bermuda, and publishing his autobiography, *The World Beneath the Sea,* in 1953 with the help of a ghostwriter, Page Cooper. His last great dream was to join Richard Leakey investigating the origins of human life in Africa, but Leakey refused his money and help because he considered Barton's methods unscientific; in fact, Barton had wanted to excavate fossils using a plow towed behind a jeep.

Barton moved around, sometimes living at the University Club in New York City, sometimes in the Willmore Hotel in Long Beach, California. For the last two decades of his long life he became a recluse at a house, which he inherited, in Cotuit, the seaside scene of his boyhood summers. He married for the first time when he was in his early sixties, but soon divorced; married again; divorced; and married for the third and last time in his eighties. He was last seen in public as an honored guest at the opening of a deep-sea exhibition at the Royal British Columbia Museum in Victoria, Canada, in 1987. Otis Barton died in 1992 and is buried in a veterans' cemetery on Long Island within view of the Atlantic.

William Beebe's vision of squadrons of manned Bathyspheres exploring the ocean never materialized, but what evolved would have pleased him beyond his wildest dreams. After the improved *Trieste II* descended to what was then believed to be the deepest place in the oceans, untethered manned submersibles took over in oceanographic research programs around the world. *Alvin,* which can carry a crew of three and dive to 14,764 feet, was launched in 1964; after thousands of hours in the abyss, it is still diving today. But will be retired in 2008 and replaced with a new submersible, which can reach depths of 21,000 feet. Other manned research and military submersibles followed, along with unmanned remotely operated vehicles—ROVs—equipped with video cameras, grappling arms, and capture devices linked by a tether to a control room on the surface, and autonomous underwater vehicles programmed to explore and return to a mother ship on their own.

The Bathysphere is still owned by the Wildlife Conservation Society, the institutional descendant of the New York Zoological Society. After the last dive in 1934, it was shipped back to the Bronx Zoo, where it remained in storage until the World's Fair in 1939 when it was the centerpiece of the society's exhibit. During the war, the Bathysphere was loaned to the U.S. Navy and taken to a testing laboratory in Maryland, where it was used to measure the effects of underwater explosions. After the war, it was returned to the society but was not on public display again until the New York Aquarium at Coney Island opened in 1957. The Bathysphere was removed from the aquarium during a renovation in 1994 and for ten years, it languished in the storage yard under the Cyclone roller coaster. At this writing, the Wildlife Conservation Society has plans to resurrect the most famous diving machine in history and again place it on exhibit at the aquarium in the spring of 2005.

John Barker, a boat builder in Bermuda, has built several full-scale fiberglass models of the Bathysphere, one of which is on display at the

headquarters of the National Geographic Society in Washington, and another at the Bermuda Underwater Exploration Institute in Hamilton. Barker also built a cutaway model of the Bathysphere, into which visitors to the institute can squeeze two at a time to experience a vicarious descent and watch the videotaped pyrotechnics of the abyssal darkness through the windows. Not many stay inside for more than a minute or two.

ACKNOWLEDGMENTS

My deepest gratitude goes to Laara Matsen, whose encouragement and advice made this book possible. I also thank Jonas Bendiksen, my friend and inspiration; Ray Troll, without whom the story would not have arrived in my life; Richard Abate at International Creative Management, who opened the door; the Sitka Center for Art and Ecology on the Oregon coast, where I was given a peaceful place to begin; and the MacDowell Colony, where I finished the work.

I owe special debts to James H. Barton, Tim M. Berra, Richard Ellis, Carol Grant Gould, and James Gould. They are among the few people other than I who have worked systematically to discover and conserve information on the lives of Otis Barton, William Beebe, Jocelyn Crane, Gloria Hollister, and the early exploration of the abyss. Their knowledge and insights were so valuable that without their help this book would be missing several important dimensions. I apologize to them for any errors of substance or nuance, for which I am solely responsible.

Without the kindness and tolerance of archivists, books like this one cannot be written. I thank Steve Johnson and Dale Boles at the Bronx Zoo Library; Diane Shapiro and the staff at the Wildlife Conservation Society photo archive; the Department of Rare Books and Special Collections at the Harvey Firestone Library, Princeton University; the attentive curators at the Bermuda National Archive in Hamilton; and the Currier Museum of Art in Manchester, New Hampshire.

I would not have been able to follow the trail of the Bathysphere expeditions in Bermuda without the help of Robert and Jean Flath of Hyper Productions; John Barker at Hole-In-the-Wall Boats; Dr. Anthony Knap, Alison Shadbolt, François Wolffe, and Gillian Hollis at the Bermuda Biological Station for Research; Jeremy Madieros on Nonsuch Island; Crystal Schultz at the Bermuda Underwater Exploration Institute; and Charlotte Andrews at the Bermuda Maritime Museum.

Many other friends and colleagues were generous with their own research, assistance, and support, including Jane Adams, Barbara Barnard, Marlene

Blessing, Mark Brinster, Gregor Cailliet, Sean Duran, Thom Holmes, Tom Jay, Aldona Jonaitis, Jane Katirgis, Ted Kheel, John McCosker, Mark Shelley, Tierney Thys, and Kerry Tremain. Kurt Esveldt and Kay Wilson contributed immensely to my life at home.

The bookmakers at Pantheon, especially the copy editors and Leah Heifferon, deserve enormous credit for their skill and patience. And thank you, thank you, Edward Kastenmeier, for taking a chance on *Descent* and helping me bring it to life.

NOTES

Beebe papers, PUL   Papers of William Beebe, Manuscripts Division, Department of
                    Rare Books and Special Collections, Princeton University Library
     WCS archives   Archives of the Wildlife Conservation Society, New York
    BBSR archives   Archives of the Bermuda Biological Station for Research, main-
                    tained at the Bermuda National Archive in Hamilton

One BARTON

The account of the events of Thanksgiving Day is from Otis Barton's *The World Beneath the Sea,* pp. 7–8, and the *New York Times,* November 25, 1926.

Barton's New York address, 20 E. Sixty-seventh Street, is printed on the letterhead of correspondence dated 1926. Other details are from my visit to the neighborhood in the spring of 2003.

Barton's family history is from interviews with his nephew, James H. Barton, in Cambridge, Massachusetts, March 2004.

The *Portrait of Master Otis Barton and His Grandfather,* a painting by William Merritt Chase, is in the collection of the Currier Museum of Art, Manchester, New Hampshire.

Otis Barton's eidetic memory and his record at the Groton School were confirmed in an interview with James H. Barton and in the unpublished notes of Francis Barton, Otis's brother, which were passed on to me by James Barton.

The descriptions of the cottage at Cotuit, Otis's construction of a pit trap, his hunting for sharks, his dive on a shark from the crosstrees of a sailboat, and his first diving helmet are from the unpublished notes of Francis Barton.

The etching of Alexander the Great on the bottom of the sea mentioned by Barton is from a thirteenth-century French manuscript, reproduced in William Beebe's *Half Mile Down,* p. 33.

The account of Barton's first helmet dive is from *The World Beneath the Sea,* pp. 1–6, and the unpublished notes of Francis Barton.

Information on New York in the 1920s is derived from the New-York Historical Society; *New York: An Illustrated History,* by Ric Burns and James Sanders; and *1927: High Tide of the 1920s,* by Gerald Leinwand.

"Beebe to Explore Ocean Bed in Tank" appeared in the *New York Times,* November 25, 1926.

Barton's devastation after reading the news of November 25, 1926, that Beebe was planning to explore the abyss is described in *The World Beneath the Sea,* pp. 7–8.

Two **BEEBE**

"Youth," by Theodore Roosevelt, is inscribed on the wall of the rotunda of the American Museum of Natural History in New York. It was assembled from four excerpts from speeches Roosevelt made at Friends School and Groton School, and his essays "America and the World War" and "American Ideals."

The description of William Beebe's Manhattan apartment is from various sources, including his diaries and an article with photographs in the *New York Sun* dated January 16, 1933, in which the writer, Elizabeth McRae Boykin, states that Beebe had lived at that address for "twenty years."

The history of the Department of Tropical Research is traced in article in the *Bulletin of the New York Zoological Society,* May–June 1938, by John Tee-Van, and in *A Gathering of Animals: An Unconventional History of the New York Zoological Society,* by William Bridges.

"Beebe didn't particularly like lecturing . . ." and information on Beebe's relationships with friends and patrons are drawn from the annual reports of the director of tropical research to the board of the New York Zoological Society, 1926–1934; interviews with Carol Gould, Beebe's biographer; and "A Personal Sketch," by Charles G. Shaw, which is included in the indispensable *William Beebe: An Annotated Bibliography,* by Tim M. Berra.

The books in Beebe's traveling library are from the *Atlantic Monthly* Bookshop list of the favorite books of famous people, circa 1938, included in the Beebe papers, PUL.

"He reported for work . . ." How Beebe spent his first pay at the New York Zoological Society and the entries from his diary for 1890, 1894, and 1899 are from Beebe papers, PUL.

". . . he was not *Dr.* Beebe at all." Beebe's advanced academic degrees and the dates honorary degrees were awarded are from author's correspondence with the administrations of Tufts and Colgate Universities in March 2003.

"Beebe published his first story . . ." Like all my citations of the chronology of Beebe's books and articles, this is taken from Tim Berra's *William Beebe: An Annotated Bibliography.*

The history and details of the *Challenger* expedition are from *The HMS* Challenger *Expedition, Admiralty Summaries of the Atlantic and Pacific Explorations;* Eric Linklater's *The Voyage of the Challenger;* and William Broad's *The Universe Below,* London, 1874.

Beebe's description of his first day at work at the Bronx Zoo is derived from an entry in his diary for October 16, 1899, Beebe papers, PUL.

Three **A DAY AT THE ZOO**

Barton's account of his discouragement on reading of Beebe's plans, his emulation of Roy Chapman Andrews, and his subsequent embarkation on the Western Asiatic Expedition with Eugene Callahan are from *The World Beneath the Sea,* pp. 7–11.

Ellen Barton's death when she was twenty-one in 1926 is an approximation drawn from an interview with James H. Barton in March 2004.

The account of Beebe's life and travels with Mary Blair, and their divorce, is from records maintained in the Washoe County, Nevada, County Courthouse; the *New York Times,* August 30, 1913; and *Two Bird Lovers* in Mexico, by William Beebe.

The quoted passage is from Beebe's *Jungle Peace,* p. 1.

The dedication to William Beebe and the quoted passage are from *Riders of the Wind,* by Elswyth Thane.

Barton describes the renewal of his plans for a diving craft in *The World Beneath the Sea,* pp. 10–11.

Barton selects Cox & Stevens Naval Architects and initiates correspondence with Irving Cox: letter (longhand) of March 10, 1928, from the WCS archives.

Barton inquires about progress on plans at Cox & Stevens from his expedition in Persia: letter (longhand) from Constantinople, July 17, 1928, WCS archives.

Barton's change of address to uptown apartment: correspondence on letterhead beginning in September 1928, WCS archives.

"Fossil Bones in a Persian Garden," by Otis Barton, *Natural History* 24, no. 2 (1929): 143–154.

Barton's early letters to Beebe ignored: *The World Beneath the Sea,* pp. 12–13.

John Butler designs and manages the building of the Bathysphere; Butler and Barton begin a yearlong correspondence: WCS archives.

American Steel Foundries refuses to guarantee the job of casting the sphere:

letter from JTR, District Sales Manager, American Steel Foundries, New York, to John Butler, December 4, 1928, WCS archives.

Letter from Butler to Barton with specifications and blueprints of design for five-ton sphere, December 18, 1928, WCS archives.

Barton's account of arranging his first meeting with William Beebe at the Bronx Zoo: *Titans of the Deep,* pp. 12–14.

The description of the Bronx, circa December 1928: Bronx County Historical Society, http://www.bronxhistoricalsociety.org/index17.html (2003).

The description of the grounds of the New York Zoological Garden, circa December 1928, and the history of the New York Zoological Society are from *Gathering of Animals,* by William Bridges.

Financial details and list of patrons are from the Annual Report of the New York Zoological Society for 1928, WCS archives.

Barton's account of his first meeting with William Beebe, quote from Beebe about air supply, and description of his office at the Bronx Zoo: *The World Beneath the Sea,* pp. 12–14.

Beebe's recollection of Theodore Roosevelt's suggestion that the diving tank be a perfect sphere is from *Half Mile Down,* by William Beebe, pp. 90–91.

Four **SPHERE**

Letters of support for Barton and his diving tank from Beebe, March 12, 1929, and from William Gregory of the American Museum of Natural History, March 11, 1929, are in the WCS archives.

Butler's selection of Watson-Stillman for machining and finish work on a steel casting to be provided by the Atlas Foundry in Buffalo, and confirmation of the joint venture between Watson-Stillman and Atlas, are from letters to Barton from Butler, April 18, 1929, and May 1, 1929, WCS archives, and a newspaper advertisement in the *Buffalo Evening News,* August 17, 1934, headlined "Beebe Reaches New Submarine Depth in Buffalo-Made Bathysphere."

Confirmation of the price of $2,300 for the finished casting of the first deep-sea tank is from a letter from William Waters to John Butler, April 27, 1929, WCS archives.

Steel casting and pattern making in 1929: from an interview with foundry owner Tom Jay, Port Townsend, Washington, March 2003.

The story of John A. Roebling and his company: Roebling Online History Archive, http://www.inventionfactory.com/history/infodesk.html (2003).

Confirmation of price, terms, and specifications for the original order of three-quarter-inch-diameter blue center wire rope: letter from John A. Roebling's Sons to John Butler, March 20, 1929.

Increase in the seven-eighths-inch-diameter wire rope from Roebling's Sons and reduction of length to 3,500 feet: letter from Butler to Barton, May 12, 1929, and description of the cable in "The Bathysphere of 1930," an article by Otis Barton included as an appendix to *Half Mile Down,* by William Beebe.

The decision to use an internal light is confirmed in correspondence with General Electric: Butler to W. L. Enfield, March 11, 1929, and C. E. Egeler to Butler, March 28, 1929, WCS archives.

Confirmation of the gift of the phone system to Barton is contained in a letter dated March 24, 1930, from John Butler, thanking C. R. Moore at Bell Telephone Laboratories, WCS archives.

The specifications for the size 14A phone and size 8A copper electrical wires remained the same for all three seasons of diving. Details are from "The Bathysphere of 1934," by John Tee-Van, Appendix C to Beebe's *Half Mile Down,* pp. 244–245; and Butler to C. R. Moore at Bell Telephone Laboratories, January 25, 1930, WCS archives.

The account of Butler's discovery of the Okonite Company, the refusal of Westinghouse and Safety Cable to bid on the job, and the rejected Simplex Cable proposal are constructed from several letters to and from Butler and the three companies in May and June 1929, WCS archives.

The specification and design of the Okonite cable is confirmed in correspondence between Butler and C. R. Moore at Bell Telephone Laboratories, dated January 25, 1930.

The design of the Bathysphere's stuffing box: "The Bathysphere of 1934," by John Tee-Van, Appendix C to Beebe's *Half Mile Down,* pp. 238–239.

Butler's discovery that ordinary glass wouldn't work at great depths is confirmed in a letter from Willard L. Morgan at Triplex Safety Glass Co. to Butler, March 15, 1929, WCS archives.

The information about water passing through glass balls used in a deep dredging operation is confirmed in a letter from H. E. Busch at the Crucible Steel Co. of America to Butler, March 15, 1929, WCS archives.

Confirmation of Butler's meeting with E. E. Free: Butler to Free, March 11, 1929, WCS archives.

The discovery of the process for making fused quartz glass by Gaudin, the general history and properties of the material, and the success of General Electric's research and development of a practical manufacturing process are described in a General Electric advertising brochure published in 1930.

E. E. Free's introduction of Butler to Jones at GE and Free's suggestion that Butler use fused quartz glass: Free to Butler, April 9, 1929, WCS archives.

General Electric agreed to supply "three clear fused quartz discs 8", in diameter, 3", thick . . . at a net price of $375 each," in a letter from W. H. Jones to Butler, May 2, 1929, WCS archives.

Butler specified 7.965 inches as the precise diameter of each fused quartz window in a letter to W. H. Jones, April 19, 1929, WCS archives.

Five BERMUDA

The dates and details of Beebe's visit to Bermuda from October 17 to November 10, 1928, during which he was offered Nonsuch Island as a field station, are confirmed in his personal diary for 1928, Beebe papers, PUL.

Beebe's discovery of helmet diving off the Galápagos Islands in 1925, his descriptions of the Kingdom of the Helmet, and his fantasies of underwater gardens are from *Half Mile Down,* pp. 66–86.

The account of Prince George's scrape with death while helmet diving with Beebe is from an entry in Beebe's personal diary for 1928, Beebe papers, PUL.

The gift of Nonsuch Island to the New York Zoological Society for use as a permanent field station is confirmed in the report of the director of the Department of Tropical Research, New York Zoological Society, for 1928, WCS archives.

Beebe's concept of a field station is from William Bridges's *Gathering of Animals,* pp. 292–305 and 463–464.

The list of patrons who supported Beebe's first expedition on Nonsuch Island in 1929 is from the report of the director of the Department of Tropical Research, NYZS, for 1929, WCS archives.

The dates and roster of the First Bermuda Oceanographic Expedition are from the report of the director of the Department of Tropical Research, NYZS, for 1929, WCS archives.

The biographical information on Gloria Hollister is from an unpublished paper written by Anthony Anable at the request of the New York Zoological Society in 1973 entitled "A Biographical Sketch of Gloria Hollister Anable," WCS archives; and Hollister's obituary in *Copeia,* fall 1988.

The love affair between Beebe and Hollister is confirmed in the translation of hieroglyphic code from his diary for 1928 at the Princeton University Library, courtesy of Beebe's biographer, Carol Gould, whose son, Grant, did the cryptography. The encoded note says: "I kissed her and she loves me." It is also obvious from many of Beebe's other entries about dancing and outings with Gloria that they were inseparable when Elswyth Thane was not in Bermuda, probably through the 1930 season. Also, in an interview, Carol Gould repeated a comment by Jocelyn Crane before her death that when Crane arrived on Nonsuch Island Hollister was "definitely the alpha female."

Barton's departure for Bermuda on June 1, 1929, is confirmed in a letter to Butler on May 21, 1929, WCS archives.

The descriptions and history of Bermuda are based on facts from W. S. Zuill's *The Story of Bermuda and Her People,* 3d ed., 1999, and several booklets, brochures, and maps published for tourists.

Barton's short stay on Nonsuch Island and his lack of ease in the laboratory are documented in Beebe's dairy for 1929, along with mentions in Beebe's report to Madison Grant for June of that year, both in Beebe papers, PUL.

In *Half Mile Down,* Beebe glosses over the miscalculation on the weight of the Bathysphere, and Barton briefly acknowledges the failure in *The World Beneath the Sea,* pp. 16 and 17. Further confirmation of the problem with the winch and the heavier sphere, and the attempts to double-rig the *Freedom* to handle the weight, are from several letters and cables between Barton and Butler, and Barton's sketch of the rigging, all from late June and early July 1929, in the WCS archive. Beebe makes no mention of Barton or the diving sphere in his diary for 1929.

Butler's correspondence with the Lidgerwood Manufacturing Company regarding the lifting power of the *Arcturus* winch, April 22 and 23, 1929, WCS archive.

Barton's two letters of June 25, 1929, and his subsequent cable to Butler on June 28 admitting that the five-ton sphere and its cable were just too heavy for the *Arcturus* winch and the *Freedom's* rigging are in the WCS archive.

Butler's letter to Barton of July 18, 1929, outlining his plan for a lighter diving sphere, is in the WCS archives.

Watson-Stillman's credit of $850 for the steel recovered from the first sphere is confirmed in a letter to Butler on August 5, 1929. Roebling's refusal to buy back the heavier seven-eighths-inch cable appears in a letter to Butler on July 12, 1929. Both are in the WCS archives.

The enormous success of the first Bermuda expedition in the summer of 1929 is documented in Beebe's monthly reports to Madison Grant at the Zoological Society, and Beebe's diary for 1929, both in Beebe papers, PUL.

Else Bostelmann's arrival on Nonsuch is confirmed in Beebe's director's report to the society for 1929 and Beebe's diary for 1929, Beebe papers, PUL.

Beebe's comments to Madison Grant on the results of trawling in a single place in the ocean are from the May 1929 monthly report, WCS archives.

Beebe's remark that begins "We sought to enter the private life of these fish" is from his article on the 1929 expedition in the *Bulletin of the New York Zoological Society,* March–April 1930, p. 47.

George Britt's article on "The Man from Mars" is from the November 11, 1929, edition of the *New York Telegram.*

Six **BATHYSPHERE**

The descriptions of Beebe's view from his veranda on Nonsuch and the lay-out of the house are derived from a visit I made to the island in the spring of 2003. Nonsuch is now a nature reserve of the Bermuda Government Parks Department located within the Castle Harbour National Park. Identifications of the various plants are based on *Living Museum: Guide to Nonsuch Island Nature Reserve,* published by the Bermuda Zoological Society in April 2001.

Beebe's ukulele playing is confirmed by his biographer, Carol Gould, in an interview in March 2003, and various other sources, including his diaries in Beebe papers, PUL.

Beebe mentions his April 17 dinner with Governor Bols and the Kiplings and his lecture on the successful 1929 expedition at the Hamilton Opera House in his May 1930 monthly report to Madison Grant, Beebe papers, PUL.

The names of Beebe's staff and the island caretakers during the 1930 season are from his monthly reports to Madison Grant, each of which included a section on personnel. Many people who are not mentioned in my narrative came and went as staff and visitors; they are omitted only to avoid bogging down the story. The main players in the drama of the first Bathysphere dives are all named in the text.

I learned that Beebe carried a copy of Murray and Hjort's *The Depths of the Ocean* on his early Bathysphere dives from Carol and Jim Gould, who have the book and were kind enough to show it to me during a visit to their home in March 2003. Beebe also included the book as one of his favorites on a list he gave to the *Atlantic Monthly.*

According to "A Personal Sketch," by Charles G. Shaw, which Tim M. Berra includes as an appendix in his *William Beebe: An Annotated Bibliography,* Beebe "particularly dislikes thinking about food. So long as it is well cooked that is all he asks and he is able to eat almost anything on earth." However, Beebe often makes reference to particularly good meals and dishes in his diary, and his personal papers include many menus from ships' dining rooms and restaurants. And according to Carol Gould, who heard it from Jocelyn Crane, Beebe ate Grape-Nuts for breakfast his entire life when he had a choice.

The sheltering of the landing on Nonsuch by sinking the *Sea Fern* as a breakwater was done late in the 1929 season and was mentioned in Beebe's annual report to the Zoological Society. Beebe also noted the improvements and Tee-Van's use of the tanks in the old ship for food fish and specimens in his *National Geographic* article, "A Round Trip to Davy Jones' Locker," published in June 1931. The remains of the *Sea Fern* still protect the landing, which is now a reinforced concrete dock set against the steep rocks.

Beebe calls Barton "Associate in Charge of Deep Diving" and a member of the Department of Tropical Research in his annual report to the society, which carries a staff list on the masthead.

Beebe writes about coining the term "Bathysphere" in *Half Mile Down,* p. 94, and Barton in *The World Beneath the Sea,* p. 27.

The arrival of the Bathysphere and Barton's success in rigging the *Ready* with the *Arcturus* winch to lift it are confirmed in Beebe's monthly report to Grant for May 1930, in Barton's *The World Beneath the Sea,* p. 28, and in Beebe's *Half Mile Down,* pp. 95–96.

The fictitious report of another successful deep-diving expedition in Hugo Gernsbacher's *Science and Invention* magazine and Barton's reassurances to Butler are from a letter from Barton to Butler, January 24, 1930, WCS archives.

The duties of the staff and crew aboard the *Ready* for launching and recovering the Bathysphere are enumerated in Beebe's *Half Mile Down,* pp. 97–98.

Beebe had one of his assistants, Virginia Ziegler, type a list of the names of the local Bermudians who worked on the *Gladisfen* and the *Ready* during the diving in 1930. They included: Gordon Virtue, R. Whiting, R. Robinson, J. Stovell, A. Smith, M. Roberts, E. Pascoe, J. Richardson, W. Sylvester, H. Bragdo, T. Green, H. Douglas, J. Bishop, E. Brown, G. Casey, C. Dowling, F. Duerden, C. Hayes, H. Paynter, A. Riley, G. H. Smith, W. Swainson, M. Thompson, H. H. Trott, R. Trott, L. Bascome, M. Bishop, C. Caisey, H. Waller, H. Whitehead, L. Baxcome, and H. Waller. The list is part of "First Log of the Bathysphere," Beebe papers, PUL.

Gloria Hollister's note on light signals and her entries on the first two test dives are from her original notebook, in the WCS archives. Details of the two test dives, including the tangled Okonite cable, are in "First Log of the Bathysphere," Beebe papers, PUL.

Seven **THE ABYSS**

Beebe had been telling reporters that he was going to dive to a half mile since the summer of 1929, probably because that was near the maximum depth the Bathysphere could attain with the 3,200 feet of cable, steel walls an inch and a quarter thick, and windows tested to just beyond that depth.

The description of the Bathysphere's atmosphere system is from Beebe's *Half Mile Down,* p. 94, and Appendix B by Otis Barton, "The Bathysphere of 1930," p. 233.

Of all the critical points at which failure would be fatal during the descents, the attachment of the cable to the Bathysphere was perhaps the most obvious. If the lifting cable broke or separated from the sphere, there was no hope of recovery. Though the Roebling clevis (the U-shaped fastening device) had been proven countless times on bridges, elevators, and other cable connections, no one had tested it under the enormous pressures and near-freezing temperatures of the abyss.

The photograph taken by Connery before the first descent is in the WCS archives.

Beebe's triangulation of his position over the cylinder and his musing about the exact location of the North Pole are from *Half Mile Down,* pp. 102–103.

Beebe's speechlessness before the first descent, the search for a seat cushion, and other details of the launching of the Bathysphere are recounted in *Half Mile Down,* pp. 103–106.

The reconstruction of the descent to 803 feet is based on Beebe's account in *Half Mile Down,* pp. 106–112, Barton's account in *The World Beneath the Sea,* pp. 29–32, and Hollister's log for dive number 4, June 6, 1930, in "First Log of the Bathysphere." The official logs typed for all the descents were edited and are slightly different from Hollister's handwritten entries in her notebooks in the WCS archives, but nothing consequential has been added or left out.

Beebe's fascination with the diminution of light is a repetitive theme in his writings and lectures about the descents. He clearly felt his observations of the transformation of color were very nearly as important as those of the creatures he saw in the depths.

Eight INFINITESIMAL ATOMS

Else Bostelmann's spectroscope, a watercolor painting of the inverted spectrum, is reproduced in *Half Mile Down* opposite p. 112.

The account of the test dive to two thousand feet on June 10, 1930, is from "Log of the Bathysphere," WCS archives.

The narrative of the aborted 250-foot dive on June 10 when the phone line failed is based on "Log of the Bathysphere" and Beebe's account in *Half Mile Down,* pp. 114–115.

Beebe reports on the fire aboard the *Ready* on the night of June 10 in *Half Mile Down,* p. 114.

The relocation of the clevis to the rear of the three holes in the lifting lug for better downward visibility is reported by Beebe in *Half Mile Down,* p. 116, and by Barton in *The World Beneath the Sea,* p. 36.

The details of the 1,426-foot descent are from Hollister's notebook, "Log of the Bathysphere," *Half Mile Down,* pp. 116–136, and *The World Beneath the Sea,* pp. 32–35.

The absence of large life-forms between 1,250 and 1,300 feet was probably the result of the Bathysphere's passing through a thermocline, a now well-understood transition layer between sun-warmed surface water and the near-zero temperatures of the abyss that presents a barrier to the creatures on both sides of the zone.

Beebe recollects thinking of Spencer's "an infinitesimal atom floating in illimitable space" in *Half Mile Down,* pp. 133–134. He laughs at himself before launching into two pages of rhapsody by writing, "The duration of all this rather maudlin comment and unnecessary philosophizing occupied possibly ten seconds of the time we spent at 1,426 feet."

Beebe's note on Hollister's sighting of a gull is from *Half Mile Down,* p. 135.

Beebe's ongoing defense of himself as an observer rather than a stuntman begins after his account of the 1,426-foot dive in *Half Mile Down,* p. 137.

The account of Hollister's dive with Tee-Van to 410 feet on June 11, her thirtieth birthday, is constructed from her article "Descending to Davy Jones' Locker," which appeared in the *Connecticut College Alumnae News* in April 1931, and a paragraph in her obituary by Tim M. Berra in *Copeia.* Beebe also mentions it in his diary from 1930: "Then Gloria and John went to 410 feet. This was her birthday present. Saw dolphin, big shark, big manta, eight shearwater feeding and fighting." Three days later, Beebe threw Hollister a party at the St. George Hotel.

"Beebe and Aide Descend 1,426 Feet Into Sea" is from the *New York Times,* June 13, 1930.

Beebe devotes an entire chapter to contour diving in *Half Mile Down,* pp. 138–145. He and Barton would repeat the dangerous shallow descents along the island shelf again in 1932 and 1934. The shallow dives are also covered in "Log of the Bathysphere." Even at two hundred feet, the Bathysphere was an observation platform in unknown waters, and Beebe was fascinated by the bottom contours that had been submerged only since the rise in water level after the relatively recent last ice age about 12,000 years before. In 1929, he wrote a chapter in *Nonsuch: Land of Water* called "Almost Island" about the undersea traces of vegetation and landforms off Nonsuch Island.

Beebe writes somewhat cryptically in his dairy about W. Reid Blair's inspection visit to Bermuda, during which he refused to come to Nonsuch Island. But in *Gathering of Animals,* William Bridges writes about Blair's skepticism and his general disdain for Beebe's work, pp. 427–429. There is no doubt that Blair lined up on the side of those who felt that the scientific value of Beebe's descents was minimal at best.

Beebe writes about the illnesses and bad luck of the expedition in July and

August in his diary, and his monthly reports to Madison Grant, both in Beebe papers, PUL. The death of Patten Jackson from appendicitis in September is documented in Beebe's monthly report and in the dedication of Jackson's Deadham, Massachusetts, school magazine *The Nobleman* to him in December 1930. In his report to Grant, Beebe wrote: "Dr. Lloyd, an old friend of the [Jackson] family and of mine arrived on October 5 and took the body back on the 7th. He went into every detail and had no blame for anyone except for the criminal statement made by a Brookline physician to the parents eight years ago, that he had removed Patten's appendix, when in reality he had only drained it."

The account of Beebe's fifty-third birthday party is from his diary, Beebe papers, PUL.

INTERLUDE, 1931

"A deck winch sputters . . ." is from a story in *Sunday World Magazine,* February 1931. The news of the quarter-mile dive played on the front pages of newspapers all over America. When it began to fade in January 1931, Beebe and Hollister embarked on separate lecture tours that were themselves covered as news, keeping the Bathysphere and the expedition in the papers and magazines through the spring.

The account of the legendary Henry Seagrave's death is from the front page of the *New York Times,* June 11, 1930.

The accounts of high-altitude ballooning and aviation records in the early 1930s are from several Internet sites, including the Smithsonian Institution Museum of Flight's http://www.centennialofflight.gov/essay/Lighter_than_ air/race_to_strato/LTA11.htm (2003).

Barton confirms that he saw *Bring 'Em Back Alive* at the Mayfair Theater on Broadway in *The World Beneath the Sea,* pp. 41–42.

Barton's gift of the Bathysphere to the New York Zoological Society with the conditions that he be part of all future diving expeditions and have exclusive rights to photography and motion pictures from the descents is documented in *The World Beneath the Sea,* p. 40, and in Madison Grant's report of the Executive Committee in the Zoological Society's annual report for 1932.

Beebe was the dominant member of the Bathysphere partnership because of his age and status as a famous naturalist and author. His relationship with Barton was cordial during the first season of diving, but in 1931 it became increasingly obvious that the two men didn't have much use for each other. They did not give joint lectures or appear in public together. In New York society, Barton was Upper East Side old money and Beebe Upper West Side artists and scientists, and their paths never crossed at parties or other events. Later, in 1933 and 1934, comments about Barton in Beebe's diaries and correspondence, and also correspondence between Thane and Edwin Conklin, confirm that Beebe neither liked nor respected Barton. Barton's negative feelings toward Beebe are also clear from letters to newspaper and magazine editors complaining about being given short shrift by his partner. James Barton told me, though, that his uncle idolized Beebe in later life.

The list of donors to the Department of Tropical Research and the Bermuda expedition are from the director's report for 1931, WCS archive.

The details of the 1931 expedition to Bermuda are from Beebe's five monthly reports, May–September, this year to Blair Niles instead of Madison Grant, Beebe papers, PUL. Beebe also summarizes the expedition in his director's report for 1931.

The death of Beebe's father and his trip back to New Jersey for the funeral are confirmed in Beebe's diary for 1931, Beebe papers, PUL.

In his director's report, Beebe cites the weather and Barton's "inability to come to Bermuda" as the reasons there was no Bathysphere diving that year. However, from Beebe's diary and the monthly reports to Niles, it is clear that other factors, including Beebe's ill health and equipment breakdowns, were equally to blame.

The visits by Edwin Chance and other potential donors to Nonsuch Island are confirmed in Beebe's diary and the monthly reports, Beebe papers, PUL.

Mrs. Chance's analysis of Beebe's handwriting is word for word from Beebe's diary for 1931, Beebe papers, PUL. His entry begins: "Mrs. Chance sent me a hand writing reading off a note I sent them. Hee hee! But I wish it were true."

Chance's gift of half the money for a season's diving is confirmed in Beebe's diary, Beebe papers, PUL, and his director's report for 1932.

Nine RENAISSANCE

Beebe's move to the Bermuda Biological Station for Research (BBSR) is documented in Chapter 4, "Beebe and His Bathysphere," in *The First Century: A History of the Bermuda Biological Station,* published by BBSR in 2003 to commemorate its one hundredth anniversary, pp. 23–31; and in the report of the director of the Department of Tropical Research, 1932, WCS archives.

Barton's arrival in Bermuda with his personal assistant, Laurance Fifilik, is confirmed in an entry in Beebe's diary for 1932, at the Princeton University

Library, and in "Second Log of the Bathysphere," kept by Gloria Hollister and edited by Beebe, Beebe papers, PUL.

The description of Beebe and Barton and their similar tonsure is based on a photograph by John Tee-Van in the WCS archives.

The description of the removal of the Bathysphere from its shed on the Darrell and Meyer wharf and Beebe's thoughts on seeing the craft after two years of storage are from *Half Mile Down,* pp. 146–147, and a photograph of the shed and wharf on p. 149.

The rigging of the *Freedom* under the direction of James Sylvester is detailed in *Half Mile Down,* pp. 147–149, and "Second Log of the Bathysphere."

Beebe's presence in Bermuda for the sole purpose of the National Broadcasting Company radio dive is confirmed in *Half Mile Down,* pp. 176–180. The arrival of the NBC crew on September 5 is noted in "Second Log of the Bathysphere," Beebe papers, PUL.

The dramatic decline in contributions to Beebe's Department of Tropical Research and the Bermuda expedition are documented in his director's report for 1932 WCS archive.

Beebe reports on his cruise/expedition to the Caribbean aboard the *Antares* with Hollister and the Chances in his director's report for 1932 WCS archive.

Beebe's illness in the spring of 1932 is documented in a letter from Edwin Conklin dated May 26, 1932, which begins "I am very sorry to learn of your prolonged illness," and in two letters from Elswyth Thane Beebe dated March 21 and May 16, in which she assures Conklin that the Bermuda expedition is still on despite Beebe's condition: "he is still in bed and can't even read with comfort and doesn't want to eat" (May 16). All these letters are in BBSR archives.

The biographical information on Edwin Conklin is from *The First Century,* pp. 13–16.

Beebe's collection of amphioxi for Conklin: Beebe's diary for 1932, Beebe papers, PUL; and a handwritten note accompanying a shipment of specimens to Conklin in November 1931, in the BBSR archives.

Conklin's offer of the directorship of the biological station to Beebe: Beebe's diary for 1928 PUL.

Beebe's letters to Conklin on January 28 (two), February 24, March 14, and May 1 confirm that he was making arrangements to set up his headquarters at the biological station, all in the BBSR archives.

The bitter relationship between Beebe and John F. G. Wheeler is still the stuff of repeated legend around BBSR, and the chapter on Beebe in the station's centennial history, *The First Century,* somewhat tactlessly begins, "In the early 1930s, a good deal of publicity, viewed by Dr. Wheeler as too sensational, was brought to BBSR by the work of a man who pioneered underwater exploration."

The terse note in which Wheeler tells Beebe he will not have the Northwest Cottage and acknowledges receipt of the plankton samples without so much as a thank you is dated February 17, 1932, and is in the BBSR archives.

Wheeler's note to Conklin telling him that Beebe would do no good for the reputation of the station is in *The First Century,* p. 24.

Beebe's note to Conklin on February 24, 1932, beginning "I do hope I have not been indiscreet," was written on the same piece of plain unlined paper as Wheeler's February 17 note denying Beebe the Northwest Cottage. The notes are in the BBSR archives.

The quotations from Olive Earle, Henry Bigelow, Goodwin Gosling, and Edward Mark concerning Beebe's presence at the station, and the references to him as "the camel," are from *The First Century,* pp. 26–28.

The account of Beebe's trip to California and his return to New York with a pail full of tide pool specimens for his wife is from a letter from Thane to Conklin, dated March 21, 1932, in the BBSR archives.

Conklin tells Beebe about the decision of the executive board of the biological station meeting in New Haven to allow Beebe to use the station in a letter dated March 14, 1932, in the BBSR archives.

The account of the solar eclipse and the flock of migrating plovers is from Beebe's diary, Beebe papers, PUL, and *Half Mile Down,* p. 149.

Barton reports on the installation of the third window in *The World Beneath the Sea,* p. 44. This is confirmed in "Second Log of the Bathysphere," Beebe papers, PUL.

Ten DISASTER

The arrival of the National Broadcasting Company crew and the names of its members are noted in "Second Log of the Bathysphere," Beebe papers, PUL, and in *Half Mile Down,* pp. 176–180.

The history of radio and its pioneers is from two Internet sites: United States Early Radio History, http://EarlyRadioHistory.us/index.html (2003); and The History of Radio, http://home.luna.nl/~arjan-muil/radio/history.html (2003).

The hurricane of September 7–11 is chronicled in *Half Mile Down,* p. 150, and "Second Log of the Bathysphere," Beebe papers, PUL.

The shark-oil barometer is still a common tourist trinket in Bermuda, and though it isn't as reliable as an aneroid barometer, some mariners swear by it.

The *Freedom* leaks so badly on September 12 that Sylvester runs for the harbor: "Second Log of the Bathysphere," Beebe papers, PUL.

The disaster of September 13, when the Bathysphere returned from three thousand feet full of water and the hatch bolt was fired across the deck, is documented in *Half Mile Down,* pp. 153–155; "Second Log of the Bathysphere," Beebe papers, PUL; and film footage of the explosive decompression shot by Laurance Fifilik and Barton, which was transferred to videotape at the Bermuda Underwater Exploration Institute.

McElrath's cable to NBC in New York advising them that the dive was still on after more than a week of waiting out the hurricane and breakdowns is in Beebe papers, PUL.

Hollister's story appeared in the *New York Times,* September 14, 1932.

Eleven ON THE AIR FROM BERMUDA

Beebe mentions the weather days and the two test dives of September 17 in his diary for 1932, Beebe papers, PUL, in *Half Mile Down,* pp. 150–156, and, with Hollister, who is quoted here, in "Second Log of the Bathysphere," Beebe papers, PUL.

The account of the Renaud family of Bridgeport, Connecticut, listening to the NBC radio dive was passed on to me by my mother, Mae, who was the oldest of the three children in the scene I construct at their home on Montgomery Street. Henry, who died in 1946, and his wife, also named Mae, who died in 1989, were my grandparents. I have thickened the description from firsthand memory of the house they lived in and family stories about my grandfather's curiosity for the exploits of Beebe and other explorers and adventurers. My mother, Mae, died in January 2004.

"BEEBE WILL TALK FROM SEA FLOOR" is from page 1 of the *Bridgeport Post,* September 6, 1932.

Unfortunately, no recording of the NBC broadcast remains, but Ford Bond's typed script is preserved among Beebe's papers at PUL. The script is quoted in its entirety at the appropriate points in the account of the descent to 2,200 feet in this chapter and Chapter 12.

Additional accounts of the broadcast, including the list of cities in the United States and England that carried it, are from "The Bathysphere Broadcast," by Gloria Hollister, which appeared in *Bulletin of the New York Zoological Society* 35, no. 5 (September–October 1932). That number was devoted entirely to the 1932 Bathysphere expedition and also carried an article by Beebe erroneously titled "A Half-Mile Dive in the Bathysphere," an account of the contour dives, and illustrated reports on hatchetfish, flying squids, and black sea dragons.

Twelve VOICES FROM THE DEEP

The weather for the NBC dive was marginal and deteriorating when Beebe made the decision to go ahead. He writes about the hurried preparations to launch and the rough conditions during the descent in *Half Mile Down,* pp. 156–175, taking his details from Hollister's log of the dive, Beebe papers, PUL number 20, in which the weather and rough sea are noted. Hollister's log of the dive is quoted verbatim from 525 feet to 1,500 feet, and then from 1,500 to 2,200 during the second half of the radio broadcast.

The transcript of Ford Bond's narration of the second half of the NBC broadcast, Beebe papers, PUL is quoted verbatim.

Beebe's sighting of "two elongate fish . . . at least six feet long," which he later described as *Bathysphaera intacta,* is recorded in Hollister's dive note for 2,100 feet during the ascent from the deepest point of the NBC dive, three minutes after the broadcast had ended. He also refers to these fish in his telegram (quoted in Hollister's log) to the Zoological Society announcing the success of the dive. In *Half Mile Down,* pp. 172–173, Beebe describes the sighting of the fish as "the most exciting experience of the whole dive." His sight descriptions of a new species flared into controversy three months later when he published his discovery of the untouchable Bathysphere fish in the *Bulletin of the New York Zoological Society* and included it in his report to the National Academy of Sciences in November 1932.

Thirteen ANIMALS FAINTLY SEEN

The telegrams of congratulations after the NBC dive are in Beebe papers, PUL.

"Fish All Lit Up . . ." is from page one of the *New York Daily News,* September 23, 1932, as reported by the Associated Press, WCS archives.

Beebe published his proposal for identification of *Bathysphaera intacta* in *Bulletin of the New York Zoological Society* 35, no. 5 (September–October 1932): 175–177.

The accounts and observations of the remaining six dives of the 1932 season are from "Second Log of the Bathysphere" and Hollister's detailed logs of dives 21–26, Beebe papers, PUL.

Beebe's observations and descriptions of the contour diving of 1932 are from

Half Mile Down, pp. 139–145, and from Hollister's detailed logs, some of which
are quoted verbatim, Beebe papers, PUL.

INTERLUDE, 1933

The history and details of the Century of Progress International Exposition are from the Internet site of the Chicago Public Library, http://www.
chipublib.org/004chicago/timeline/centuryprog.html (2003); and the Internet site of the Illinois State Library, http://www.museum.state.il.us/exhibits/
athome/1920/welcome.htm (2003).

The details of Piccard's *Century of Progress* balloon are from the Internet site
of the Chicago Museum of Science and Industry, http://www.msichicago.org/
exhibit/transport/piccard.html (2003).

The criticism by Carl Hubbs and John Nichols of the sight description of
Bathysphaera intacta almost certainly began with Beebe's presentation of the new
fish and its name in the *Bulletin of the New York Zoological Society.* Their comments appeared in print, however, some time later, Hubbs's in "Reviews and
Comments," *Copeia: A Journal of Cold Blooded Vertebrates,* July 1935, p. 105; and
six months later, Nichols's in *Natural History,* January 1936, pp. 88–89.

The display of the Bathysphere at the American Museum of Natural History in New York is confirmed in the report of the director of the Department
of Tropical Research for 1933, WCS archives.

Barton's attendance at the opening of the exhibit at the American Museum
is confirmed in a photograph by an unknown photographer, WCS archives.

Barton's cigarette card still exists in collections, and can be seen on the
Official William Beebe Website, http://hometown.aol.com/chines\6930/mwl/
beebe1.htm (2003).

The accounts of Barton's trips to the Bahamas and Panama with his hired
film crews are from *The World Beneath the Sea,* pp. 45–68.

The history of underwater cinematography and the Williamson brothers is
from *Take Me Under the Sea: The Dream Merchants of the Deep,* by Thomas
Burgess, pp. 163–243, with details drawn from photographs of the Williamsons
and their Photosphere between pp. 116 and 117. This book is a well-documented
survey of pioneer undersea writers, photographers, and artists and a must-read
for any ocean enthusiast and lover of the bizarre.

The history of Louis Boutan and the first underwater still photographs is
from an Internet site, "Photography 101: Part 1, Of Underwater Photography
History," by Michel Gilbert and Danielle Alary: http://divemar.com/diver-
mag/archives/dec96/gilbert1_dec96.html (2003).

Beebe's second expedition aboard the *Antares* with Hollister and the Chances, and the quoted poem by T. G. Aspenwall, are from Beebe's diary, June 14–30, 1933, Beebe papers, PUL, and from the report of the director of the Department of Tropical Research for 1933, WCS archive. The results of the expedition, including a list of specimens, appeared in the *Bulletin of the New York Zoological Society,* July–August 1933.

Beebe's retrieval of the Bathysphere from Chicago and his comments on its dilapidated condition are from *Half Mile Down,* pp. 181–182.

Fourteen GROSVENOR AND THE *GEOGRAPHIC*

The accounts of the founding of the National Geographic Society and its magazine and the life of Gilbert Grosvenor are from C. D. B. Bryan's *The National Geographic Society: 100 Years of Adventure and Discovery,* the only authorized history of the society, pp. 24–49.

The founding members of the National Geographic Society were Charles Bell, banker; Israel C. Russell, geologist; Commodore George W. Melville; Frank Baker, anatomist; W. B. Powell, educator; General A. W. Greely, polar explorer; Grove Karl Gilbert, geologist; John Wesley Powell, geologist; Gardiner Greene Hubbard; Henry Gannet, geographer; William H. Dall, naturalist; Edward E. Hayden, meteorologist; Herbert G. Ogden, topographer; Arthur P. Davis, engineer; Gilbert Thompson, topographer; Marcus Baker, cartographer; George Kennan, explorer; James Howard Gore, educator; O. H. Tittmann, geodesist; Henry W. Henshaw, naturalist; George Brown Goode, naturalist; Cleveland Abbe, meteorologist; Commander John R. Bartlett; Henry Mitchell, engineer; Robert Muldrow II, geologist; Commander Winfield S. Schley; Captain C. E. Dutton; W. D. Johnson, topographer; James C. Welling, educator; C. Hart Merriam, chief of the U.S. Biological Survey; Captain Rogers Birnie, Jr.; A. H. Thompson, geographer; and Samuel S. Gannett, geographer. The list is from *The National Geographic Society: 100 Years of Adventure and Discovery,* p. 25.

Grosvenor's letter to Beebe, December 12, 1933, offering to accept a proposal for sponsorship of ocean exploration in 1934, Beebe's letter of proposal on December 20, 1933, Beebe's budget proposal, Grosvenor's telegram awarding the $10,000 grant, and his follow-up letter are from Beebe papers, PUL.

Fifteen NEW EYES

Beebe's publications in 1933 are from Tim Berra's *William Beebe: An Annotated Bibliography,* and the 1933 report of the director, Department of Tropical Research, New York Zoological Society WCS archive.

Barton's letters to newspapers about Beebe grabbing the limelight are referred to in an entry in Beebe's diary, July 31, 1934, Beebe papers, PUL.

Beebe's receipt of the first payment on his grant from the National Geographic Society is confirmed in a letter dated March 5, 1934, Beebe papers, PUL.

Barton's continuing failure to film shark attacks is described in *The World Beneath the Sea,* pp. 69–70.

Beebe's telegram notifying Barton of the National Geographic Society grant and his intention to dive in the summer of 1934 are from Beebe papers, PUL.

Barton's move to the West Side of Manhattan is established from a return address in correspondence with John Butler, May 26, 1934, in WCS archives.

Beebe's quoted observations of the Bathysphere at the Watson-Stillman shop and the failure of the windows in a pressure test are from Beebe's *Half Mile Down,* pp. 181–182.

Beebe's correspondence with P. K. Devers at General Electric and the A. D. Jones Optical Works, letters dated April 17, 1934, and April 16, 1934, respectively, are in Beebe papers, PUL.

C. E. Adams of the Air Reduction Company agreed to give his system to Beebe at cost in return for mentions and photographs in publicity in a letter dated March 28, 1934, Beebe papers, PUL.

A description of the atmosphere and power systems by John Tee-Van is included in "The Bathysphere of 1934," Appendix C to *Half Mile Down.*

Accounts of the departure from New York, the arrival in Bermuda, and the staff of the expedition are from Beebe's diary for 1934, Beebe papers, PUL; and the 1934 report of the director of the Department of Tropical Research WCS archive.

Barton's account of turning up the pressure during the test at Watson-Stillman is from *The World Beneath the Sea,* p. 70.

The exchange of letters between Barton and Butler, May 26, 1934, and June 3, 1934, is from Bathysphere correspondence, WCS archives.

Sixteen **BLOTCHES OF LIGHT ON A BLACK FIELD**

The date the Bathysphere was shipped from New York aboard the *Queen of Bermuda,* July 3, 1934, is confirmed in "Third Log of the Bathysphere," Beebe papers, PUL.

Details of the *Queen of Bermuda* are from "Bermuda Queen Recalled," by William H. Miller, on the Web site of the Ocean Liner Museum, http://www.oceanliner.org (2003).

The account of Beebe's confrontation with Barton on the bow of the tug and his quoted comment are from Beebe's diary, 1934, Beebe papers, PUL.

The expedition motto "Three hours and a half mile" is from Beebe's *Half Mile Down*, p. 187.

The publicity campaign the National Geographic Society was building around Beebe's descents is documented in Beebe's correspondence with Grosvenor, LaGorce, and Fisher, and telegrams, Beebe papers, PUL.

Beebe's exchange of cables with Franklin Fisher regarding good wishes for Kepner and Stevens in their balloon ascent are in Beebe papers, PUL.

The account of the stratosphere balloon record attempt by Kepner, Stevens, and Anderson is from C. D. B. Bryan's *The National Geographic Society: 100 Years of Adventure and Discovery*, pp. 355–360.

The accounts of E. John Long's good relationships with the members of the Beebe expedition and his fondness for Jocelyn Crane are drawn from references in correspondence between Long and Beebe after Long returned to New York in late August 1934, Beebe papers, PUL.

The account of the preparations for diving at the Darrell and Meyer wharf and Sylvester's demand for a dress rehearsal before going to sea are from *Half Mile Down*, pp. 186–190, and the article of the same name that appeared in the December 1934 issue of *National Geographic* magazine, p. 669.

The account of Beebe's decision to cancel the next day's test dive on the evening of August 6, his change of mind on the morning of August 7, and Barton's race in the canoe to catch the tug and barge is from *The World Beneath the Sea*, pp. 71–73. This differs slightly from Beebe's own record in *Half Mile Down*, p. 190, in which he writes that he made the decision to dive on the night of August 6, rather than on the morning of August 7 as implied in the Barton account.

Barton's account of the results of the August 7 camera dive is from *The World Beneath the Sea*, p. 74.

Seventeen WORLD RECORDS

The account of the August 11 dive to 2,510 feet is assembled from "The Third Log of the Bathysphere," from the WCS archives; Beebe's *Half Mile Down*, pp. 199–214 and Appendix F, "Unedited Telephone Conversations on Dive Number Thirty," by Beebe and Gloria Hollister, pp. 264–279; Beebe's diary for 1934, Beebe papers, PUL; and Barton's *The World Beneath the Sea*, pp. 75–77.

The scene in which John Long jokes about "world records" is a composite drawn from references in his letters to Beebe in August and September 1934, found among Beebe's correspondence with Long, Beebe papers, PUL.

Beebe's attempt to find a spectrograph and his concentration on the physics of light are documented in an exchange of letters between Beebe and E. O. Hulbert, an optical physicist at the Naval Research Laboratory, Beebe papers, PUL.

The account of testing the light at 1,900 feet and again at 2,000 feet by holding fingers in front of the window is from a story by Otis Barton that appeared in the *New York Herald Tribune,* November 11, 1934, and *The World Beneath the Sea,* p. 76.

Beebe's observation and description of the pallid sailfin is from *Half Mile Down,* pp. 204–206; the November–December 1934 issue of the *Bulletin of the New York Zoological Society,* pp. 190–191; and "Unedited Telephone Observations on Dive Number Thirty," by Beebe and Hollister.

Beebe's observation and description of the three-starred anglerfish is from *Half Mile Down,* p. 211; the November–December 1934 issue of the *Bulletin of the New York Zoological Society,* pp. 191–192; and "Unedited Telephone Observations on Dive Number Thirty," by Beebe and Hollister. His full description of the fish in the *Bulletin* reads:

> It was almost six inches or a little longer in length, a deep, almost regular oval in outline, black and with small eyes. Paired fins were not seen, but pectorals at least were undoubtedly present. The remainder of the webbed fins were typical for this group, dorsal four and anal four. But close in front of the dorsal were two oval, sessile bulbs like those of Cryptosparas.
>
> Half-way between the bulbs and the eye was the posterior of three tall illcia (stalks), slender, apparently stiff, each about one-third the length of the fish. The anterior one originated almost half way between the second and the eye, the two posterior ones arising closer together. The anterior was somewhat stouter but of the same length as the others. Each had a slightly enlarged tip. These tips gave out a strong, pale yellow light, powerful enough to illumine the adjacent dorsal skin when the fish was not in the direct path of my beam. The most distinct character was the mouth, which was only slightly oblique and furnished with short, even, sharp teeth, these characters setting it well apart from all known genera of Ceratiidae. This Three-starred Anglerfish remained in full view, in the darkness and in the light, for fully five seconds, long enough for me to see and fix in mind the recorded proportions and characters.

Beebe's observation and description of the five-lined constellation-fish is from *Half Mile Down,* pp. 212–213; the November–December 1934 issue of the

Bulletin of the New York Zoological Society, pp. 192–193; and "Unedited Telephone Observations on Dive Number Thirty," by Beebe and Hollister. The full description in the *Bulletin* reads:

Again the observational data is the same for this as for Bathyembryx, with a change in depth to 1,900 feet. On the way to the surface, 470 feet higher than where we had seen Bathyceratias, this fish, one of the most unexpected and gorgeous deep-sea inhabitants I have ever seen, appeared, quite as abruptly as the preceding. This too I saw both in the dark and in our light, thus showing almost all its principal charaacters. It swam slowly past and then turned broadside on, very close to the sharp boundary of the beam.

At the surface, it would have passed for a strange Chaetodon or Acanthurus, but it was assuredly far from being either, unless a distant relation wholly adapted for life at three hundred fathoms. It was almost round—I estimated about five by six inches—with long, moderately high, continuous, vertical fins and a deeply concave tail, all exactly like the fins of Acanthurus. The eye was very large and interrupted the dorsal, cephalic profile. The mouth was large, and the jaws projecting. The teeth and the pelvic fins were invisible to me, if any were present. The pectorals were rounded and rather short.

The Bathysphere swung a few degrees to starboard and left the fish in absolute darkness, and I suddenly saw the amazing beauty of the photophores. There were five rows of these, one along the median line from the eye to the peduncle. Two others divided the upper half of the body into unequal thirds, curving downward along their whole length, and two more occupied the same relative positions below the center line.

The equatorial line consisted of larger lights than in the other four lines. The chief light in each of the numerous units was bright but pale yellow, shining from a large central photophore. Each of these in turn was surrounded with a circle in the median row, or a half circle in all the other rows, of tiny, brilliant purple lights. This formed a wonderfully beautiful pattern of illumination, resembling that of no other abyssal fish I have ever seen.

Before it left, it turned and faced me, its movement being perfectly distinct, even in the darkness, from the shifting lines of lights and their gradual foreshortening. I now saw that the breadth of the fish was not great, again about as in a surgeon fish. Even in the partly front view, the sides were distinctly lighted by a soft reflection from the photophores.

I can assign this fish to no known family, but the Five-lined Constellation-fish assuredly deserves a name, pending the time when, with improved trawling nets, we will be able to bring one to the surface, alive or dead, and by dissection allot it more definitely to some known or unknown group.

The excerpt from John Long's account for the National Geographic Society that was published in its entirety in dozens of newspapers is taken from the *New York Times,* August 12, 1934.

Eighteen HALF MILE DOWN

The description of Else Bostelmann's studio is from a photograph by John Tee-Van, WCS archives.

Beebe and Bostelmann's collaboration on images of the new abyssal creatures is from *Half Mile Down,* pp. 213–214.

Beebe and his staff take two days off for helmet diving and net towing: from *Half Mile Down,* pp. 214–215; and Beebe's diary, August 13 and 14, 1934, Beebe papers, PUL.

Barton's concern about the lifting capacity of the winch and the effects of pressure on the Bathysphere and windows a half mile down is described in *The World Beneath the Sea,* pp. 77–78.

The account of the August 15 dive to 3,028 feet is assembled from "The Third Log of the Bathysphere," WCS archives; *Half Mile Down,* pp. 215–226 and Appendix F, "Unedited Telephone Conversations on Dive Number Thirty-two," by Beebe and Gloria Hollister, pp. 264–279; Gloria Hollister's notebook for the 1934 season, WCS archives; Beebe's diary for 1934, Beebe papers, PUL; and Barton's *The World Beneath the Sea,* pp. 78–79.

John Knudsen's photograph of Beebe and Barton after the 3,028-foot descent is from the WCS archives.

Epilogue TITANS OF THE DEEP

First epigraph: Many modern oceanographers and marine biologists, including Robert Ballard, Sylvia Earle, John McCosker, and Bruce Robison, cite Beebe, Barton, and the Bathysphere as inspiration for their own decisions to explore the abyss. Those who have seen the Bathysphere or a model of it are awed by the courage it took to descend in the four-and-a-half-foot steel ball at the end of a cable.

Second epigraph: William Gregory, curator of living and extinct fish at the American Museum of Natural History and one of the earliest supporters of

Barton's dream of deep-ocean exploration, wrote this in an article for *Nature* after seeing the film shot in 1934. Barton later quoted him in *The World Beneath the Sea*, p. 74.

The summary of the final dives of the Bathysphere by Beebe, Barton, Tee-Van, Crane, Hollister, Bass, and Ramsey gives short shrift to what were astonishing experiences. Beebe doesn't mention them in *Half Mile Down,* probably because, as in this narrative, the 3,028-foot dive is the climax of his story. The details of the final descents are drawn from Beebe's diary and "Third Log of the Bathysphere," compiled from Hollister's notebooks, Beebe papers, PUL.

"Beebe and Barton . . . never spoke to each other again." The evidence of their total estrangement is in the absence of any collaborative work or article after the summer of 1934, but also in comments made to Jim and Carol Gould by Jocelyn Crane, which were passed on to me in an interview with the Goulds in the spring of 2003.

Beebe asked Long to approach LaGorce and Grosvenor for extra money for the 1934 season in a letter dated September 24, 1934, Beebe papers, PUL. Beebe kept up a correspondence with John Long into 1936. In a letter dated October 2, 1934, Long accepted Beebe's offer to make a Bathysphere descent if there was another diving expedition and took a shot at Barton. "I am extremely flattered by your invitation, left-handed though it was, to join you in a bathysphere dive," Long wrote. "I can assure you that I have no photographic ambitions and never write letters to the editor of the New York Times. If Dr. G. fails [to provide more funds] you can come down to Washington and sit in on the New Deal! Science needs a leader! I'll gladly handle the publicity pro bono publico."

Barton's account of finishing his film *Titans of the Deep* is from *The World Beneath the Sea,* pp. 80–127. Though he recounts, through his ghostwriter Page Cooper, the adventure of moviemaking off Panama, much of the narrative focuses on the menagerie of jungle cats and snakes he brought back to New York, some of which became house pets in his apartment. Even in a contrived and not entirely reliable autobiography, Barton's eccentricity comes through, and it's easy to see why the stolid Beebe, whose life depended on hard work and a rigid set of judgments of himself and others, couldn't stand his partner.

The positive review of *Titans of the Deep* from *Film Daily,* November 14, 1938, is pure Hollywood flackery, typical of the era when a few studios, including Grand National, distributed all the movies and controlled this kind of review.

Beebe's letters to the editor denying any involvement with *Titans of the Deep* appeared in *Science* magazine and other publications in April 1939.

Carl Hubbs was the snide *Copeia* reviewer of *Half Mile Down* in July 1935.

He devotes more than half the review to "animals faintly seen," but also reports that Beebe "by his magic pen, whirls us from the origin of life on earth to the First Wonderer; then the Thinker; finally on through Columbus and Magellan to the explorers of the submarine waters, all climaxed of course by Beebe's own half mile descent."

The controversy surrounding Beebe's sight descriptions of the untouchable Bathysphere fish, the pallid sailfin, the three-starred anglerfish, and the five-lined constellation-fish continue to break the surface of ichthyological debate. In an unpublished article, "The Cryptozoological Fishes of William Beebe," the distinguished marine writer and artist Richard Ellis writes:

ABSTRACT: During his dives in the bathysphere during the early 1930s, oceanographer and science writer William Beebe (1877–1962) identified many new species of deep-sea fishes and invertebrates. Most of these were described from specimens that had been trawled up and preserved, but there were several for which no specimens were collected, and which were only seen by Beebe through the portholes of the bathysphere. In these cases, we have no specimens to confirm the existence of Beebe's sightings, only his "eyewitness" accounts and the drawings that accompanied these descriptions. No one since Beebe has ever seen or described these species, and there is a possibility that Beebe fabricated them.

Buck Peterson, pilot at the Monterey Bay Research Institute of the remotely operated vehicle *Tiburon,* which dives beyond sunlight many days every year, also told me that catching a glimpse of any species of anglerfish is rare, so it's easy to be skeptical about Beebe's sightings.

In 2001, though, Erich Hoyt, senior research associate with the Whale and Dolphin Conservation Society in England, gave Beebe the most definite benefit of the doubt in his book *Creatures of the Deep* (Buffalo, N.Y.: Firefly Books, pp. 44–45). "Most of [Beebe's] descriptions fit creatures we know today inhabit the middle layers, just where he saw them," Hoyt wrote. "Beebe's 'untouchable bathysphere fish' turned out to be a new species of dragonfish. Other fish described by Beebe, such as the viperfish and the little devilfish, can be positively identified through his illustrations . . ."

Beebe's departure from the sea, his eventual establishment of his home and field station at Simla, and his enduring relationship with Jocelyn Crane are recounted in William Bridges's *A Gathering of Animals,* pp. 463–465. In the mid-1940s, Bridges reports, Beebe told Crane to "shop for a jungle," and it is she who came up with the idea for the base in Trinidad.

Fairfield Osborn, known as Fair, was the son of Henry Fairfield Osborn, one of the founders of the Zoological Society and Beebe's first patron there. Fair Osborn's obituary of Beebe, from which this quote is taken, is from *Animal Kingdom*, the successor publication of the Zoological Society *Bulletin*, August 1962.

Richard Welker did a good job of pulling together much of the evidence of Beebe's role as a pioneer ecologist in *Natural Man: The Life of William Beebe* in 1974. That book was criticized by the surviving members of the Bermuda expeditions, who at the time included Jocelyn Crane and Gloria Hollister, both of whom reportedly felt that Welker cast Beebe in an unfavorable light as a womanizer, opportunist, and egotist who did, in fact, quash Barton's early career in favor of his own. The new biography by Carol Gould (Nov. 2004, Islands Press) should go the full distance in establishing Beebe as a true ecologist and one of the founders of that discipline as a legitimate science.

Rachel Carson not only dedicated *The Sea Around Us* to Beebe, but frequently mentioned him in lectures as one of her greatest inspirations and teachers.

The account of Gloria Hollister's life after the Bathysphere expeditions is from the biographical sketch written for the records of the Wildlife Conservation Society in 1974 by her husband, Anthony Anable, and her obituary in *Copeia* in 1988.

The summary of the rest of the lives of John Tee-Van and Else Bostelmann is from Bridges's *A Gathering of Animals*, pp. 305, 391–392, 430, 458, 462, 477, 480, 482, and 497.

The accounts of Barton's dives for *Life* magazine in 1952 are from *The World Beneath the Sea* and an article, "Oh, to See in the Deep," by oceanographer Andreas B. Rechnitzer, then a graduate student at Scripps Oceanographic Institute in La Jolla, California, who was aboard the research vessel *E. W. Scripps* as an observer during the dives. The article was anthologized in *The Spark in the Sea: Adventures in Marine Science,* edited by David J. Horrigan (Home Planet Books in association with the San Diego Section of the Marine Technology Society, 2003).

The details of the rest of Otis Barton's life are from *The World Beneath the Sea* and conversations in 2003 between the author; James H. Barton; oceanographer John McCosker, who talked with Barton; and marine biologist Gregor Caillet, who worked on a deep-sea trawl with Barton in Monterey, California, in 1974. A letter on the stationery of the Willmore Hotel is particularly telling in terms of Barton's eccentricity and his relationship with Beebe and is worth quoting here in its entirety. The occasion was a request for a remembrance to be included in a testimonial album commemorating Beebe's retirement from the Zoological Society on his seventy-fifth birthday in 1952. Barton wrote in longhand, with some orthographical slips:

WILLMORE HOTEL

ROOMS AND APARTMENTS IN A REFINED ATMOSPHERE

315 WEST THIRD STREET

LONG BEACH 12, CALIFORNIA

I have read and enjoyed some ten of Dr. Beebes books. Let's hope
writers of animal stories will be more and more taxonomists and com-
paratie anatoomists. With the glorious days of large scale jungle and big
game adventure nearing extinction, we are gladen by the realization that
the same kind of adventure lurks in any tropical garden, or in any garden
or back yard. As Theodore Roosevelt said we must learn to perceive not
meerly to see the animals and parts.

Otis Barton

COFFEE SHOP SPECIALIZING IN FOOD YOU WILL LIKE

AT PRICES YOU CAN PAY

BIBLIOGRAPHY

Ballard, Robert, with Will Hively. *The Eternal Darkness: A Personal History of Deep-Sea Exploration.* Princeton: Princeton University Press, 2000.

Barton, Otis, with Page Cooper. *The World Beneath the Sea: The Story of the Deepest Dive Ever Made by Man.* New York: Thomas Y. Crowell, 1953.

Beard, Daniel Carter. *The American Boy's Handy Book: What to Do and How to Do It.* Boston: David R. Godine, 1882.

Beebe, William. *The Arcturus Adventure.* New York: Putnam, 1926.

———. *Beneath Tropic Seas.* New York: Putnam, 1928.

———. *Half Mile Down.* New York: Harcourt, Brace, 1934.

———. *Jungle Peace.* New York: Holt, 1918.

———. *Nonsuch: Land of Water.* New York: National Travel Club, 1932.

———. *Two Bird Lovers in Mexico.* New York: Houghton Mifflin, 1905.

———. *Zaca Venture.* New York: Harcourt, Brace, 1938.

Berra, Tim M. *William Beebe: An Annotated Bibliography.* Hamden, Conn.: Archon Books, 1977.

Bridges, William. *A Gathering of Animals: An Unconventional History of the New York Zoological Society.* New York: Harper & Row, 1974.

Broad, William. *The Universe Below: Discovering the Secrets of the Deep Sea.* New York: Simon & Schuster, 1997.

Bryan, C.D.B. *The National Geographic Society: 100 Years of Adventure and Discovery.* New York: Abrams, 1987.

Burgess, Thomas N. *Take Me Under the Sea: The Dream Merchants of the Deep.* Salem, Ore.: Ocean Archives, 1994.

Burns, Ric, and James Sanders. *New York: An Illustrated History.* New York: Knopf, 1999.

Carroll, Lewis. (Charles Lutwidge Dodgson.) *Alice's Adventures in Wonderland* and *Through the Looking-Glass.* London: Penguin, 1997.

Doyle, Arthur Conan. *The Lost World.* Oxford: Oxford University Press, 1995.

Earle, Sylvia. *Sea Change: A Message of the Oceans.* New York: Putnam, 1995.

Hoyt, Erich. *Creatures of the Deep.* Buffalo, N.Y.: Firefly Books, 2001.

Leinwand, Gerald. *1927: High Tide of the 1920s.* New York: Four Walls Eight Windows, 2001.

Linklater, Eric. *The Voyage of the Challenger.* New York: Doubleday, 1972.

Mosley, H. N. *Notes by a Naturalist: An Account of Observations Made During the Voyage of H.M.S. "Challenger" Round the World in the Years 1872–1876.* Rev. Ed. New York: Putnam, 1892.

Murray, John, and Johan Hjort. *The Depths of the Ocean: A General Account of the Modern Science of Oceanography Based Largely on the Scientific Researches of the Steamer* Michael Sars *in the North Atlantic.* London: Macmillan, 1912.

Robison, Bruce, and Judith Connor. *The Deep Sea.* Monterey, Calif.: Monterey Bay Aquarium Press, 1999.

Sterrer, Wolfgang. *Bermuda's Marine Life.* Flatts, Bermuda: Bermuda Zoological Society, 1992.

Thane, Elswyth. *Riders of the Wind.* New York: Frederick A. Stokes, 1926.

Welker, Robert Henry. *Natural Man: The Life of William Beebe.* Bloomington: Indiana University Press, 1975.

Zuill, W. S. *The Story of Bermuda and Her People.* 3d ed. Oxford: Macmillan, 1999.

INDEX